WASHINGTON PUBLIC SHORE GUIDE
Marine Waters

James W. Scott
Public Access Program Manager

Melly A. Reuling
Research Writer

Don Bales
Cartographer

State of Washington
Booth Gardner, Governor

Department of Ecology
Andrea Beatty Riniker, Director

UNIVERSITY OF WASHINGTON PRESS
Seattle   London

The preparation of this document was financially aided through a grant to the Washington Department of Ecology with funds obtained from the National Oceanic and Atmospheric Administration, and appropriated for Section 306 of the Coastal Zone Management Act of 1972.

Copyright © 1986 by the University of Washington Press
Printed in the United States of America

All rights reserved. No part of this publication may be reproduced or transmitted in any form or by any means, electronic or mechanical, including photocopy, recording, or any information storage or retrieval system, without permission in writing from the publisher.

Library of Congress Cataloging-in-Publication Data
Scott, James William
　Washington public shore guide.

　Bibliography: p.
　Includes index.
　1. Shore-lines—Washington (State)—Guide-books.
2. Coasts—Washington (State)—Guide-books. I. Reuling, Melly A.　II. Bales, Don.　III. Title
GB458.8.S46　1986　　　917.97'0443　　　85-40976
ISBN 0-295-96334-4
ISBN 0-295-96335-2 (pbk.)

PREFACE

Washington's 2,400 mile saltwater shoreline is a center of activity in the northwest. Its beauty and wide variety of recreational opportunities attract many people, and its diverse ecosystems support extensive plant and animal life. The complex nature of our shore poses tough and sometimes conflicting management problems: how does the "public" get adequate use of this resource with out impinging on the rights of private property owners? how does the natural community survive the pressure of enthusiastic human use? and how can legitimate needs for commercial, industrial and residential uses be satisfied without unduly conflicting with the former?

Managing the shore in fair and environmentally sound ways and increasing opportunities for public access are responsibilities that we and our associates face daily in administering the Shoreline Management Act. Since the act's passage by the State Legislature, 14 years ago, significant progress has been made in meeting its objectives. In fact, many of the sites in this guide would not exist without these efforts. Some recent accomplishments made by the Department of Ecology include marking public sites with the new public access logo, improving permit procedures relating to shoreline management, and publishing this guide.

The progress made by the Department of Ecology only supplements the fine work done by many other agencies in providing public access sites and in managing the state's natural resources. Most shoreline recreation opportunities are provided by the Department of Natural Resources, with its public beaches program, and the State Parks and Recreation Commission, with its numerous state parks. Not to be overlooked are the smaller programs of the departments of Game and Fisheries and the sites provided by the federal Fish and Wildlife Service and the National Park Service.

Also the areas provided by many local government agencies are important components of the total public access program. All these agencies are instrumental in protecting the natural features of the shoreline; in fact, natural resource management programs in existence today are much stronger than at any previous time in our history.

Nevertheless, statistics prove that in spite of these efforts the public still has few recreational opportunities along the shore and many of the shore's natural resources continue to be damaged and destroyed. Along many stretches of shore there are few places where the public can go. Access is often blocked by private ownership and "no trespassing" signs, and the public areas that are available are often overcrowded. The nearly 700 sites in this book represent only about 425 miles of public shoreline, or 21% of the shore (excluding the outer coast which is mostly public). Since only half of that has upland access, the public only has real access to about one-tenth of the inland saltwater shoreline. Of this, some is hardly known about or used at all.

We can only do so much. With the right of shoreline use and access there goes a public responsibility, a responsibility to protect and preserve the resources we enjoy here in the northwest, and also a responsibility to respect private property rights.

With the publication of this guide, we hope that you will not only become more aware of access opportunities, but also gain a greater understanding and appreciation of the natural environment of the shoreline. We hope that this guide will enhance your enjoyment of our saltwater shore, encourage its conservation, and help you understand the resource management problems we face. If so, our efforts in putting it together have been worthwhile.

James W. Scott
Melly A. Reuling
Olympia, Washington

WASHINGTON PUBLIC SHORE GUIDE

MARINE WATERS

NOTICE:
The authors have made every attempt to be as thorough and accurate as possible and have based this guide on public records and/or published sources, but inasmuch as beach ownership boundaries are often disputed and rarely accurately surveyed and because beaches and headlands are inherently dangerous places to visit, users of this guide are advised that neither the authors nor the publisher make any warranty as to the accuracy of the data or the safety of beach areas for the public to visit.

CONTENTS

SITE INFORMATION

WHATCOM COUNTY	31
SKAGIT COUNTY	45
SAN JUAN COUNTY	66
ISLAND COUNTY	122
SNOHOMISH COUNTY	147
KING COUNTY	159
KITSAP COUNTY	182
PIERCE COUNTY	205
THURSTON COUNTY	227
MASON COUNTY	234
CLALLAM COUNTY	250
JEFFERSON COUNTY	270
GRAYS HARBOR COUNTY	290
PACIFIC COUNTY	304
WAHKIAKUM COUNTY	323

RESOURCE INFORMATION

PUBLIC ACCESS LOGO	9
WASHINGTON'S SHORELINE MANAGEMENT	10
A WORD ABOUT CONSERVATION	12
HISTORY	14
HISTORY OF PUBLIC BEACHES	20
DEPT. OF NATURAL RESOURCES BEACHES	21
STATE PARKS	22
NATIONAL WILDLIFE REFUGES	24
HABITAT MANAGEMENT AREAS	25
KEY TO SITE INFORMATION	26
COUNTIES INCLUDED IN THIS GUIDE	28
SEALS	34
RED TIDE	42
PADILLA BAY	48
ESTUARIES	60

RESOURCE INFORMATION CONT.

SAN JUAN NAT'L. WILDLIFE REFUGE	70	BEACH HIKING	272
ORCAS	104	RAIN FORESTS	276
WASHINGTON FORTS	142	SHOREBIRDS	300
FLOATS	156	BOWERMAN BASIN	302
WASHINGTON STATE FERRIES	176	DUNES	306
CRABBING	198	WILLAPA NAT'L WILDLIFE REFUGE	316
TIDES	216	BEACHES	326
NISQUALLY NAT'L. WILDLIFE REFUGE	233	ACKNOWLEDGEMENTS	338
CLAM DIGGING	246	ABOUT THE AUTHORS	339
SHELLFISH	247	SELECTED REFERENCES	340
SPITS	262	INDEX	342
DUNGENESS NAT'L. WILDLIFE REFUGE	266		

INTRODUCTION

The Washington Public Shore Guide serves two main purposes: to provide public information on shoreline access, and to increase public understanding and appreciation of the shoreline environment. The access part of the guide tells you where to go on the coast, how to get there, and what facilities are available at a particular site. The environmental part describes seashore life found in and out of the water, outlines the physical and geological processes that make a beach, and shows how these forces fit together to form diverse coastal ecosystems.

Washington's 2,421 mile marine coastline is one of the most varied in the nation. The south half of the outer Pacific Coast is characterized by broad sandy beaches, dunes, and prolific wetlands. This wide coastal plain has the most popular recreational beaches in the state. Further north, on the outer Olympic Peninsula, the beaches are narrow and rocky and often backed by high forested bluffs. The northern region is sparsely populated and because there are no large estuaries, good harbors, or industrial sites it remains largely unaltered; this area is ideal for beach hikes or camping trips.

In contrast to the outer Pacific Coast, the coast of Puget Sound is distinguished by quiet rocky shores, broad mudflats, and protected harbors. Greater Puget Sound is a complex system of interconnected inlets, bays, and channels with tidal seawater entering from the west, and cold freshwater streams entering from the surrounding mountains. Greater Puget Sound includes the Straits of Georgia and Juan de Fuca, Hood Canal, and the San Juan Islands. These areas provide a variety of environmental and recreational values.

The marine influence on western Washington keeps winters mild and summers moderately cool. The Olympic and Cascade mountain ranges rake the moisture out of the ocean breezes, causing large variations in precipitation from area to area. Up to 200 inches of rain per year may fall on the western slope of the Olympic Peninsula and a more moderate 35 to 40 inches per year falls in the Puget Sound lowlands between the two ranges. Rainfall is lightest in July and August; this is also the time that recreation along the coast is most popular.

Fisheries in Washington attract thousands of sport fishers every year and are also a major commercial industry. Puget Sound waters are rich in nutrients and support a wide variety of marine fish and shellfish species. Washington is most renowned for its salmon; all five species of Pacific salmon spend part of their lives in the salt waters of Puget Sound and the Pacific Ocean before returning to spawn in their streams of origin.

Washington is one of the most important shellfish producing states in the nation. Its six species of clam make clam digging a major reason people visit the beaches. Geoducks, perhaps the most unusual shellfish in Washington, are the world's largest burrowing clams, and have recently become the most significant commercial clam fishery in the state. These huge clams can weigh up to twenty pounds, and have been known to live 130 years. Oysters abound as well; Willapa Bay is the major oyster producing area on the West Coast, with oyster production inside Puget Sound not far behind.

The wealth of marine fish and shellfish, plus the varied habitat along the coast, support the rich diversity of marine birds and mammals found in the region. The sound is an important resting place, feeding area, and wintering ground for thousands of birds migrating on the Pacific Flyway. Waterfowl dot the calm inland waters, while shorebirds race up and down with the ocean surf and great blue herons stalk the salt marshes. Eagles, hawks, and falcons soaring overhead, feed on the smaller birds and fish. There is no doubt that Puget Sound and the outer coast can keep any bird watcher happy.

The most common marine mammals seen along the saltwater shoreline are harbor seals. Their insatiable curiosity often brings them very close to humans on shore or in boats. In the northern part of the sound, elephant seals and several species of sea lions can be seen at breeding rookeries or lounging on haulout areas at low tide. Three species of whales are found off the coast and in Puget Sound. Playful killer whales are often seen around the San Juan Islands and in the straits, where dolphins and porpoises are also residents. Mammals that live along the shore include river otters, beavers, raccoons, and mink, among many others.

The same qualities that attract marine life to Washington waters attract thousands of people to the coast every year. As the population and interest in outdoor recreation increase, pressures on the natural environment also increase. While this guide aims to promote use and enjoyment of the coast, it is also designed to increase understanding of, and respect for, the unique and often fragile coastal environment. Just as we are able to enjoy the coast today, our awareness and conservation will enable future generations to enjoy it also.

PUBLIC ACCESS LOGO

This logo has been adopted by the Department of Ecology to mark public access sites. It was designed in 1984 by Gina M. Forth, a student at Shorecrest High School in Seattle. Gina was the winner of an art contest sponsored by the Department of Ecology to design a public access symbol.

There are two versions of this logo. The sign shown here marks where the public has a right of access. The other sign entitled "SHORE VIEW" marks viewpoints where the water can be seen but can not be reached. Watch for the signs. Many of the sites in this book are marked, and some small sites, that are not in this guide are also marked as are sites on lakes and streams which are also not in this saltwater guide.

The Department of Ecology would like to get all public access sites along the shore marked, but we have no way of knowing if signs are in place, have be removed or otherwise vandalized. You can help insure that public access area signs will be installed and replaced. If you know of a public area that does not have signs please contact your local planning department, your local park department or the Shorelands Program of the Department of Ecology, Mail Stop PV-11, Olympia, WA. 98504.

WASHINGTON'S SHORELINE MANAGEMENT

During the late 1960s, a series of conflicts over environmental issues erupted in the state of Washington. Many of the most serious conflicts were related to the use and abuse of the state's invaluable water resources and adjacent shorelines. Thermal and industrial waste pollution, depletion of fisheries, offshore oil drilling, oil spills, and the loss of beaches to residential and commercial development all became the focus of intense controversy.

Spurred by this controversy, the legislature created the Department of Ecology in 1970. The department was organized to carry out the state's new responsibilities in environmental management. The legislature declared it a state policy that

> "...it is a fundamental and inalienable right of the people of the state of Washington to live in a healthful and pleasurable environment and to benefit from the proper development and use of its natural resources."[RCW43.21C.020(3)]

A year later the legislature went further and passed the State Environmental Policy Act of 1971 (SEPA). SEPA is primarily a disclosure statute that requires environmentally sound planning, and the airing of issues involved in governmental decision making, to allow increased public scrutiny of proposed actions. The legislature declared *"that each person has a fundamental and inalienable right to a healthful environment and that each person has a responsibility to contribute to the preservation and enhancement of the environment."* Today, SEPA remains the state's most comprehensive statement of environmental policy.

The events which led to passage of SEPA convinced Legislators that the environmental rights of the people of the state were most seriously jeopardized by misuse of state's water and shoreline resources. In 1971 the legislature also passed the Shoreline Management Act (SMA).

The SMA is designed to manage the land 200 feet inland from shorelines and associated wetlands. The act resulted in a massive state program to manage the state's shoreline resources with a new attention to the environmental, economic, and social impact of resource utilization. It places most of the technical administration in the hands of local governments, while the state maintains a watchdog role.

In the act, the legislature declared that the Department of Ecology will give preference to shoreline uses in the following order:

1. Recognize and protect the statewide interest over local interest;
2. Preserve the natural character of the shoreline;
3. Result in long-term over short-term benefit;
4. Protect the resources and ecology of the shoreline;
5. Increase public access to publicly owned areas of the shoreline;
6. Increase recreational opportunities for the public in the shoreline;
7. Provide for any other element deemed appropriate or necessary.

In the implementation of the shoreline act, the public's opportunity to enjoy the physical and aesthetic qualities of natural shorelines is to remain a high priority.

To follow these directives, the Department of Ecology has recently placed a greater emphasis on public access to shorelines. Access along the state's shorelines is not limited to public parks, but is provided in many forms. It is required to go with commercial developments such as restaurants and condominiums, as well as with industrial developments that are on the waterfront. At these sites public access may be anything from a simple viewpoint, to a fishing pier or picnic facility.

To foster public awareness of shoreline management the Department of Ecology has prepared this book to guide you to areas along the state's saltwater shoreline that have been set aside for public use and enjoyment. In addition, the Department of Ecology is providing standardized signs to local agencies so public sites along the shore can be marked. This guide and the signs should facilitate enjoyment of the state's shoreline.

Beach ownership and shoreline development are constantly changing. The information on public sites in this guide is complete to the best of DOE's knowledge as of summer 1985. The DOE will make an effort to keep beach access sites marked and shoreline information up to date. Any information from those of you who are using the public access sites will greatly aid the department in its effort to preserve your right to the state's shoreline.

A WORD ABOUT CONSERVATION

A high priority of shoreline management in Washington is preservation of the natural character of the shore and protection of its environment by controlling shoreline development. Promotion of public access and recreational opportunities along the shore are also priorities. Reaching these goals at the same time requires that users of the beach be aware of their impact on natural areas and be conscientious about preserving the shoreline for future visits and future generations.

This book has been written to guide you to all types of public areas along the shore: commercially developed areas, developed parks, and areas that are still in their natural state. Because these areas belong to you, they should be treated as though they were in your front yard. Please be conscientious about disposing the garbage that you bring in, and contribute to maintaining the area by picking up any litter you find.

The areas included in this book which still remain in their natural state need special care to preserve the delicately balanced shoreline environment. Since these areas are already protected from commercial and industrial development, any damage to the environment comes from the visitors. Although most of us who enjoy natural areas have an interest in preserving the environment, some people are not aware of the effects their actions may have.

Continued freedom to experience our natural environment requires a willingness to assume responsibility for our actions. Rather than impose strict, and therefore limiting rules to protect the environment, the Department of Ecology would like to help you use natural areas in environmentally sound ways. The following guidelines are recommended for everyone who enjoys Washington's natural shores.

1. Restore the environment to its original state after making physical changes (such as moving rocks, building fires, or digging) so the next visitor can have the same natural experience you had.

2. If you dig clams or other animals, be sure to fill in the holes that you make. Piles of mud may suffocate other burrowing species, and holes left open may leave clams exposed to air for too long.

3. Be careful not to move or crush marine life while exploring tidepools or turning over rocks. Leaving the underside of a rock exposed will kill any organisms attached to it. Moving an organism in a tidepool just a foot could alter its exposure to tidal cycles enough so that it dies.

4. Don't collect or disturb plants and animals in areas designated by law as biological preserves. If for some reason you need to do specific work in these regions, apply to the managing agencies for permission. In San Juan County, for example, the marine fauna and flora are protected by law. Permission to collect biological material may be obtained in advance from the Director of Friday Harbor Laboratories of the University of Washington.

5. Obey fish and game laws with respect to season, bag limits, and size and sex of the animals taken for food. These laws have been developed to strike a balance between human use of each species, and maintaining a healthy population.

6. Do not dig up or catch more animals than you can use, whether it be for study or for food.

7. If you are using animals for study, do as much studying as possible at the shore. If animals must be taken to a lab, return them to their natural habitat or save them for future use.

8. Make use of flora and fauna found on manmade structures such as docks, piers and floats to take some pressure off natural areas.

9. Avoid coming to close, or otherwise bothering marine mammals. The Federal Marine Mammal Protection Act strictly prohibits the disturbance of any marine mammal and prohibits the possession of marine mammals or marine mammal products such as whale bones.

If you are involved in teaching shoreline ecology, or just enjoying the shore with friends, encourage those you are with to follow conservation practices. By spreading the word, you will help keep our beaches enjoyable for generations to come.

To understand shoreline environments and the associated human impacts, a discussion of shoreline ecology is included at the end of the book under "BEACHES". Whether you wish to understand these environments or not, you should respect the area you are visiting.

HISTORY

The coast of Washington has existed in its present form about 25 million years, only a flash of time geologically. Over the ages, uplifting and folding of the earth's crust has formed mountain ranges; lava spewed from volcanoes has created an abundance of minerals and rich soil; glaciers have carved valleys; and rainfall has eroded and filled the valleys to form rivers, lakes and Puget Sound. The high rock cliffs, wide sandy beaches, estuaries and numerous inlets we enjoy today reflect the varied geologic forces that have shaped the area where the land of Washington meets the water.

The Puget Sound area as we know it now has only existed since the Pleistocene glacial ice age, about 13,000 years ago. During the Pleistocene epoch huge sheets of ice advanced and retreated, carving the rock in their path. At its densest point the ice was 4,000 feet thick (seven times the height of the Space Needle) over the Seattle area. When the ice finally retreated to the north it left behind deeply carved valleys which eventually filled with water to form the north-and-south oriented bays and passages which now make up greater Puget Sound.

After the basic arrangement of the coast was made by glaciers, water began designing the finishing touches. Sediments were key to the creation of today's coastline and these were supplied in enormous quantity by the glaciers of the Pleistocene era. Waves and currents reworked the glacial sediment, molding landforms in the valleys like frosting on a cake. The results are the beaches that now surround Washington waters.

Because Puget Sound beaches were originally formed glacially, their composition is often a mix of mud, sand, gravel, and boulders. It is typical for glaciers to deposit sediments in a random fashion like this. Flowing water, on the other hand, sorts and deposits sediments by size, resulting in a more uniform beach composition. Since the original glacial deposits, the effects of wind, waves, and currents have reshaped some beaches along the coast to form shoreline of more consistent composition.

Humans have been living on these beaches for at least 10,000 years. It is not known how the first people got to Washington, although most anthropologists feel that Native Americans originally migrated from Asia. During the ice ages bands of people may have walked across a then-existing land bridge between Russia and Alaska. It is theorized that they then moved down the coast to settle in more moderate climates.

The earliest record of human life on the coast of Washington is that of the coastal Indians. Several distinct but similar cultures lived along the coast, all of them getting most of their livelihood from the shore and the water. Because the resources of the region were so abundant, Northwest Coast Indians were the most affluent of all North American native cultures.

The staples of the native economy were salmon and cedar. Salmon were caught as they entered the rivers for their spawning run, then smoked and dried for use later in the season. Their diet was supplemented with seafood and shellfish, waterfowl, small game, roots and berries, and out on the Olympic Peninsula the Makah tribe hunted whales.

The versatile cedar tree was used to make spacious longhouses, storage boxes, clothing, utensils, baskets and a variety of skillfully made canoes.

Washington State Historical Society, Tacoma, WA photo

The water was used as both a highway and a hunting ground by the Indians. Their canoes, some of which were fifty feet long and carried up to thirty people, varied in design with their uses. There were different canoes for hunting, traveling and war parties, as well as for the open ocean, rivers, and protected waters.

A typical native community would consist of a group of longhouses along the shoreline, often near the mouth of a river. Each longhouse would be occupied by several related families. A record of family descent was often carved on totem poles or a representation of lineage was painted on boards. Wealth and status in a village were hereditary, although no one could lead the rest of the community except by example and force of character.

Because the Coastal Indians had a strong belief in spirits of nature, the natural resources used by Indians were treated with respect and appreciation. Indians believed that the animals they hunted made a voluntary sacrifice to sustain their brother humans. Salmon were believed to be people who lived in longhouses under the ocean. They transformed themselves into fish and swam to the rivers every spring for the Indians to harvest. If they were not treated with respect, they would not come back. Each tribe performed a First Salmon Ceremony with the first catch of the year, greeting the Salmon People on their return from the ocean.

For thousands of years, Northwest Coast Indians lived on the shorelines without upsetting the ecology or depleting the resources they depended on. When European explorers came to claim the territory and exploit the resources of the Northwest, both the natural environment and indian life were changed forever. Although white men originally came looking for gold, silver and new trade routes it was, ironically, fur pelts provided by the Indians that first made this a region coveted by many nations.

The first known explorers in the region were searching for the Northwest Passage, the mythical water route across the North American continent that could save trading ships the long and dangerous journey around the Horn of South America. By the late 1700s ships from several countries had reached the Northwest Coast. Russian and Asian expeditions traveled down the North Pacific coast. Spanish ships arrived from their colonial outposts in Mexico, leaving a reminder of their presence in the places they named: San Juan Island, Fidalgo Island, Rosario Strait. French and British explorers and representatives of the young United States also sailed to the Northwest Coast looking for trade and for new lands to claim.

The first European to discover the potential of the fur trade on the Northwest Coast (as the Russians had done farther north in Alaska) was Captain James Cook. When he reached the coast in 1778, his sailors found that the Indians would trade lush sea otter pelts for old knives and pieces of iron. When Cook's ships reached China, the pelts sold for 150 silver dollars apiece. As the word got out that there were fortunes to be made on the Northwest Coast, nations competed for control of the territory.

England sent Captain George Vancouver to explore the Pacific Coast and to claim territory for the British Empire. Vancouver spent the summer of 1793 in Puget Sound, producing remarkably accurate maps of the area. His first lieutenant, Peter Puget, led an expedition in two small boats exploring and mapping the southern end of the sound which now bears his name.

The United States was also interested in claiming the western coast of its vast continent. The most significant early American claim on the Northwest Coast was of the Columbia River, made by Captain Robert Gray in 1792. Since a nation that discovered a river could lay claim to all the lands it drained, claiming the giant Columbia had major importance.

By the beginning of the nineteenth century, England and the United States were setting the pattern that the Northwest's economy would follow for the next hundred years. First furs, then timber and seafood, were harvested to be sold elsewhere.

The Native American tradition of respect for the environment was not necessarily continued by white men. The supplies of furs, fish and timber were so abundant that it seemed no amount of harvest could deplete them. It was said that at that time a soldier could kill as many as a hundred shorebirds with a single blast from his gun.

It took only a few decades to decimate some populations. The lucrative trading of sea otter pelts to China led to the extinction of sea otters in this region. Seal and beaver populations also declined sharply because of the popularity of their pelts. Whales, hunted for their oil, were threatened by unregulated harvest. As settlement increased and industry grew, all of the Northwest's resources—from trees and animals to clean air and water—were pushed closer to their limits.

By the 1820s, only England and the United States still claimed "Oregon Country" which was the territory between Russian America to the north, Spanish America to the south, the Continental Divide to the east, and the Pacific Ocean. England's claims were increased by the presence of the Hudson's Bay Company, a great trading empire which had established posts

Washington State Historical Society, Tacoma, WA photo

Washington State Historical Society, Tacoma, WA photo

there. The U.S. claim was strengthened by the success of the Lewis and Clark expedition which traveled down the Columbia and reached the Pacific Ocean in 1805.

The United States and England signed a ten year joint occupation agreement for Oregon Country in 1818, so that both nations could settle and trade while waiting for a final resolutions of borders and ownership. The agreement was renewed until 1846, when the northern border between British and American territory was set at the 49th parallel. Two years later Oregon Country became an official U.S. territory, and in 1853, the land north of the Columbia River was granted separate status as Washington Territory. By then, settlers had reached Puget Sound and the white settlement of Washington's shorelines was well under way.

Indians felt increasingly threatened by the influx of white settlers coming to claim their land. While the American and British disputed over land, the fact that the Indians were the original owners was largely disregarded. Some Indian tribes were passive and cooperative with white men, but others decided to strike out in self-defense. Starting in the late 1840s skirmishes resulted in many deaths of settlers and Indians.

One of the primary missions of Washington's first territorial governor, Isaac Stevens, was to get the Indians to sign treaties giving up their land and to move them to reservations. Unlike many of the huge Indian reservations around the Rockies, those in the Northwest were generally smaller and located closer to the original tribal centers. There are eleven reservations along the saltwater shore of Washington.

The completion of the transcontinental railroad in 1883 brought more settlers to the Pacific Northwest and provided a more efficient way to ship freight and resources to the expanding markets of the eastern United States. Washington was granted statehood in 1889, and by the late 1890s economic growth had greatly accelerated; lumber sales boomed,

Washington State Historical Society, Tacoma, WA photo

mining flourished, and shipbuilding was becoming an important industry. The Alaska gold rush of 1897 further boosted economic prosperity in the Puget Sound region, especially in Seattle. In 1914, World War I created a demand for all Northwest resources, causing another growth spurt in Washington.

Growth and development in Washington correlated closely with use, and destruction, of natural resources. By the turn of the century the demands of loggers, fishermen, miners, industrial developers, and businessmen were challenging the environment's ability to maintain itself. It became apparent to some that both the natural beauty of the state as well as the foundations of its economy were being depleted.

Resource management and conservation practices began developing in the early 1900's in response to the threats to the environment. Areas of the state were set aside as national parks, national forests, wilderness areas, recreation areas, and state parks. In the 1930s, during the Great Depression, Civilian Conservation Corps (CCC) work projects developed these areas for public use.

Washington State Historical Society, Tacoma, WA photo

Growth picked up again in the 1940s as cheap hydroelectric power from the new Columbia River dams attracted many large industries. This placed even greater pressure on the resources, but at the same time, environmental consciousness was also growing. By the 1960s pressures of the environmental movement were beginning to counteract the pressures of environmental degradation. There was increasing interest in minimizing pollution, protecting natural areas and maintaining a high quality of life.

Now, development and the conservation movement are growing simultaneously in Washington. The balance between economic activity and the preservation of the resource-rich environment needs constant attention. Maintaining this balance, while striving to maintain the high quality of life found in the Pacific Northwest, is one of biggest challenges in Washington's future.

Washington State Historical Society, Tacoma, WA photo

HISTORY OF PUBLIC BEACHES

When the original thirteen American colonies became independent and self-governing, the founders adopted land use laws based on the English legal system. This meant that the shores and beds of navigable waters were considered property of the colonies. Upon admission to the union, each new state retained these rights. Washington was no exception. When the Washington State Constitution was adopted November 11, 1889, the state asserted ownership of the beds and shores of navigable waters up to and including the line of ordinary high water (mean high water).

At that time, all the tidelands of the state were publicly owned. However, Washington's state constitution contains no provision allowing upland property owners any rights of access to saltwater for transportation, fish and shellfish, propagation, or other water-oriented industry.

To correct this situation and provide revenue for the state, the 1889-1890 legislature authorized the sale of public tidelands to private individuals. During the ensuing years, approximately 60 percent of Washington's state-owned beaches were sold before the practice was discontinued in 1971. Unfortunately, most Puget Sound beaches were sold during this time.

Today 1,300 miles of saltwater tidelands are state-owned. About 300 miles of beach, including Pacific coastal beaches, are managed by the State Parks and Recreation Commission or the Department of Game and Fisheries. About 1,000 miles of state beaches are managed by the Department of Natural Resources (DNR).

DEPARTMENT OF NATURAL RESOURCES BEACHES

The Department of Natural Resources (DNR) manages public trust lands with a multiple use concept. The beaches are managed to produce income for the trustees, although some are most suitable for non-income purposes such as recreation. Therefore, the DNR has inventoried state-owned tidelands and identified those tidal beaches best suited for reserves, commercial use, and public use. Approximately 75 percent of the DNR managed tidelands were identified as candidates for public use beaches. The DNR beaches included in this book are the most usable public tidelands in that inventory.

Often DNR beaches are not associated with any easily identified upland features and the boundaries of a beach can be difficult to locate. Throughout much of Puget Sound, public tidelands were marked by four-foot tall white posts. It is difficult to keep these markers in place, however, particularly on exposed windswept beaches. The DNR is evaluating alternative methods to physically mark the boundaries of otherwise unidentifiable beaches. While using these beaches, be aware that the uplands are usually private and most of the beaches are bordered by private land. Please be respectful of private property, and if any doubt exists as to where a beach boundary is, stay out. To help with identification, the DNR has published a series of guides called "Your Public Beaches," which have aerial photos and maps of many of its beaches. The guides, as well as additional information about public tidelands, are available from the DNR:

Department of Natural Resources
Division of Marine Land Management
Olympia, WA 98504
(206)753-5324

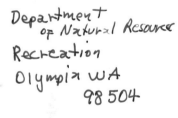

Department of Natural Resource Recreation Olympia WA 98504

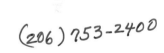

(206) 753-2400

STATE PARKS

The Washington State Parks and Recreation Commission manages much of the public land along the saltwater shoreline. Included under the management of State Parks is the outer coast (under the Seashore Conservation Act), many large camping areas, scattered picnic areas and boat launches, interpretive centers, historic sites, and extensive marine facilities.

Camping Information:

Most of the campsite areas along the saltwater shoreline are managed by State Parks. They offer three types of campsites: standard, utility, and primitive. Standard campsites, which are the most common, have picnic tables, nearby running water, sink-water drains, garbage disposal and flush comfort stations. Almost all parks with standard campsites have coin operated showers. Utility campsites have all these facilities, plus individual hook-ups for domestic water, sewer, and/or electricity. Primitive sites are not provided with nearby flush stations and may not have all of the amenities of a standard campsite. A fee is charged for overnight camping and varies according to the type of site.

Seasons:

From April 1 through September 30, all state parks in western Washington are open daily from 6:30 a.m. to dusk.

Most parks are open all year, although some are closed from October 1 to March 31. Check with State Parks if you want to camp during that period. For parks remaining open, winter hours are 8 a.m. to dusk.

Rules:

You can stay a maximum of ten consecutive days in any one park.

Most parks have some sites that can accommodate both a vehicle and a trailer up to a combined length of 32 feet. Because the sites at each park vary, check with specific parks to make certain there are sites large enough for your recreational vehicle.

All pets MUST be on an 8-foot controlled leash at all times.

There is an extra charge per night for a second vehicle, unless that vehicle is hitched to a motor home.

Some state parks offer group reservations for both day use and overnight camping. Write State Parks for details.

Most state parks are available on a first-come, first-serve basis. State parks that accept reservations for the period between Memorial Day and Labor Day are Belfair, Fort Canby, Twin Harbors/Grayland Beach, and Moran. Reservations can be made only by sending a reservation application directly to the park. Reservations must be made at least fourteen days in advance of your first requested camping date and may be made as early as January. Call or write State Parks for a reservation application.

Fort Worden State Park, near Port Townsend, has its own year-round reservation system for full hook-up campsites, conference facilities and vacation housing. Write State Parks for a brochure or call (206) 385-4730 for more details.

Marine facilities:

Washington State Parks offers many marine facilities for boaters. Fees are charged for use of the moorage facilities at certain marine parks. The fees are applicable May 1 through Labor Day between 3 p.m. and 8 a.m. Most of these are collected on a self-registration basis.

For additional information on State Parks contact:

Washington State Parks and Recreation Commission
7150 Cleanwater Lane, KY-11
Olympia, WA 98504
(206) 753-2027

Toll free within Washington,
1-800-562-0990 summers only

NATIONAL WILDLIFE REFUGES

Dwindling and damaged habitat has seriously reduced some wildlife and fish populations. To help stem this trend and preserve our fish and wildlife heritage, the U.S. Fish and Wildlife Service manages a network of National Wildlife Refuges (NWR).

There are nine refuge areas along Washington's saltwater shoreline. Refuges at Willapa Bay, Dungeness Spit, and Nisqually Delta are accessible by land, while Flattery Rocks, Quillayute Needles, Copalis, Protection Island, and San Juan Island refuges are all islands that are, for the most part, not accessible.

The refuges are important to provide the necessary habitat—food, water, cover, and space—to ensure the survival of birds, mammals, reptiles, amphibians, fish, and endangered plants and insects.

Abundance and type of wildlife varies with the seasons at most refuges. The greatest concentrations of ducks and geese on wetland refuges in the region are present from fall to spring, and especially during the fall and spring migration periods.

Most refuges welcome visitors. Many of them are excellent places to view and photograph birds and other wildlife, and to learn about wildlife and habitat management.

Refuges permit recreational uses such as hunting and fishing when they are consistent with the primary wildlife purposes. All uses are subject to state and federal regulations, and the refuge manager should be contacted to obtain specific information. Camping and picnicking facilities are not usually provided.

For further information on National Wildlife Refuges contact:

U.S. Fish and Wildlife Service
Lloyd 500 Building, Suite 1692
500 NE Multnomah Street
Portland, OR 97232
(503)231-6828

HABITAT MANAGEMENT AREAS

The Department of Game owns more than 20,000 acres of natural habitat and public access areas within the coastal zone. Management of these areas is directed toward improving habitat for all wildlife and the varied needs of outdoor people who are attracted by the wildlife.

There are four major Habitat Management Areas along Washington's coastal zone: the Oyhut, Johns River, Skagit, and Nisqually. In addition, the Department of Game operates five waterfowl areas, six saltwater access areas and holds the title to several thousand acres of tidelands in Skagit Bay.

To use Department of Game areas a Conservation License is required. The licenses can be purchased at any store that sells fishing or hunting licenses.

Regulations and information pertaining to each area, such as road use and camping facilities, may be obtained from the Washington Department of Game.

Washington Department of Game
600 N. Capitol Way, GJ-11
Olympia, WA 98504
(206) 753-5700

KEY TO SITE INFORMATION

Information about public shore sites is arranged in the Guide by county and further subdivided by shore segment. Maps are included to provide the general location of each site, but for the most part you should obtain local street maps, or nautical charts if you are a boater, to guide you. A matrix table is provided for each shore segment which lists characteristics about each site. Where a dot appears the characteristic exists at the site. The word "undev" after a sitename indicates the area is an undeveloped property with no facilities for public use. These are usually recently acquired parcels of land which are intended to be developed in the future. The following definitions apply for the characteristics listed in the matrix:

ACRES: The number of acres of contiguous ownership at the site regardless of the portion which has be developed for public use. A "NA" indicates that acreage figures were not available. Some acreages are not given such as for DNR beaches which are only tidelands and not readily measurable in terms of acres.

CAMP UNITS: The number of individual camping sites. Often these areas include picnic tables and recreational vehicle hook-ups.

PICNIC UNITS: The number of picnic tables without associated campsites.

RESTROOMS: A dot indicates public restrooms are available, but they may range from crude pit toilets to elaborate flush facilities.

FIREPITS: This means there are designated places to build campfires. Often these are pedestal type, charcoal only, grills. In most instances you must bring your own firewood, as wood is not provided and gathering is not allowed.

SWIMMING BEACH: The area has a beach suitable for swimming which may or may not be supervised by a lifeguard.

BOAT LAUNCH (lanes): The number of launching lanes, whether paved or not. Most of the lanes are paved.

BOAT MOORAGE (slips/buoys): The number of moorage slips, buoys, and pier spaces where boats can be tied. This includes transient and permanent spaces. No marinas are included that do not provide transient moorage.

PUBLIC PIER: This is an area that has a public pier extending into the water which is suitable for fishing.

DRINKING WATER: Drinking water available.

VIEWPOINT: This means there is either a view from the shoreline to some scenic vista or the site is an upland area providing a view of the shoreline and water.

SHORELINE LENGTH (feet): The total length of the shoreline of contiguous ownership regardless of how much of the shoreline is usable beach.

ROCK BEACH: A beach which is mostly composed of an outcropping of solid bedrock.

SAND BEACH: A beach which is mostly composed of sand-sized particles.

GRAVEL BEACH: A beach which is mostly composed of gravel-sized particles. Often these beaches are a mix of mud, gravel, and small boulders.

MUD BEACH: A beach mostly made up of mud. Mud beaches are usually difficult if not impossible to walk on.

SAND DUNES: These are usually upland from the beach where sand particles are deposited by the wind to form undulating topography.

TIDE POOLS: These are usually found in association with rock beaches where pools of saltwater are trapped when the tide retreats. The pools form microhabitats for a variety of marine life.

WETLANDS: Wetland is a term that refers to a broad lowland area, usually consisting of mud and marsh, that is inundated regularly by the tide. Wetlands feature special vegetation that is adapted to a wet, boggy environment. Wetlands usually are associated with fresh water and occur at estuaries.

BLUFFS: These are formations rising abruptly from the tideline which often eliminate any form of usable beach at high tide. They may be composed of solid rock, or compacted soil and gravel as in many glacial deposited formations found in Washington State. They often offer great views of the water.

COUNTIES INCLUDED IN THIS GUIDE

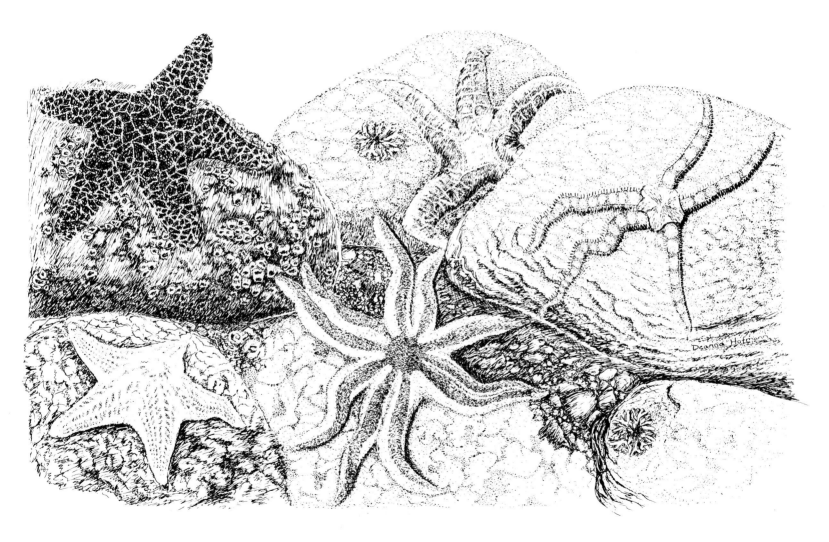

Drawing by Deanna Hofmann

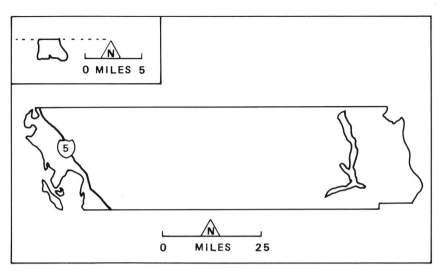

WHATCOM COUNTY

Whatcom County, which shares its northern border with Canada, has 106 miles of saltwater shoreline. There are several islands within the county's boundary, the largest of which is Lummi Island. Point Roberts, a small piece of U.S. territory at the southern tip of a Canadian peninsula, is the westernmost part of the county. Its sandy beaches and high feeder bluffs are accessible by boat or by driving down through Canada.

Bellingham is the largest city in the county and the fifth largest in Washington. It is a center for fish processing, wood pulp and paper production, oil refining, agriculture, boat manufacturing, and higher education. Mount Baker towers over the city to the east, and the San Juan Islands are scattered along the straits west of Bellingham Bay.

Being the major gateway to Canada, (Vancouver, British Columbia is just forty miles to the north), Bellingham is a popular tourist area. It is a popular spot for boaters, offering easy access to the San Juans, Victoria, B.C., and the Canadian Gulf Islands. There are ample boating facilities at Blaine, Sandy Point, and Bellingham.

Interesting shoreline characteristics of Whatcom County include Semiahmoo Spit near Blaine, which has a newly developed county park; the Lummi Peninsula, which is connected to Portage Island at low tide, and Portage Island itself (permission for access must be obtained from the Lummi Tribal Office). Lummi Island, accessible by ferry, has both forested mountains and pastoral farmland and offers beautiful views of Mount Baker and the San Juan Islands.

A drive south out of Bellingham on Chuckanut Drive will take you past numerous high viewpoints overlooking Bellingham Bay, Chuckanut bay, and the San Juan Islands. Larrabee State Park is located along Chuckanut Drive and has a mixed shoreline of sandy beaches and rocky bluffs, and is an excellent spot to watch tidepool life.

Watching the sun sink behind the San Juan Islands from the shoreline of Whatcom County is an experience that should not be missed by anyone in this part of the world.

WHATCOM COUNTY
NORTH

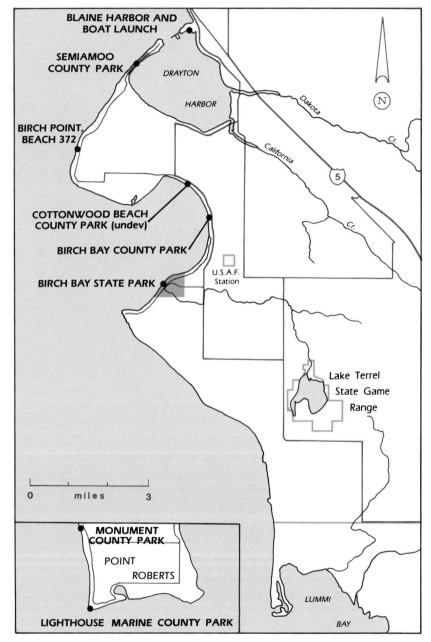
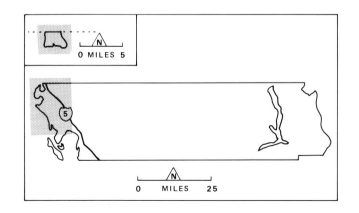

WHATCOM COUNTY
NORTH

PUBLIC SHORE

	ACRES	CAMP UNITS	PICNIC UNITS	RESTROOMS	FIREPITS	SWIMMING BEACH	BOAT LAUNCH (lanes)	BOAT MOORAGE (slips/buoys)	PUBLIC PIER	DRINKING WATER	VIEWPOINT	SHORELINE LENGTH (feet)	ROCK BEACH	SAND BEACH	GRAVEL BEACH	MUD BEACH	SAND DUNES	TIDE POOLS	WETLANDS	BLUFFS
BIRCH BAY COUNTY PARK	NA					•						1,320	•	•						
BIRCH BAY STATE PARK	191.9	167	190	•	•	•				•	•	6,000	•	•			•			
BIRCH POINT, BEACH 372	NA											2,930	•	•						
BLAINE HARBOR AND BOAT LAUNCH	60.0		•				1	520	•	•		2,300								
COTTONWOOD BEACH COUNTY PARK (undev)	14.0					•						1,200		•	•					
LIGHTHOUSE MARINE COUNTY PARK	22.0	20	28	•	•		2			•	•	3,960	•	•						
MONUMENT COUNTY PARK (undev)	8.2											500	•	•						
SEMIAHMOO COUNTY PARK	322.0		10	•	•	•						6,700	•	•						

BIRCH BAY COUNTY PARK
From I-5 take the Birch Bay-Lynden Road to Birch Bay Drive, head south to the roadside park. Beach only. Parking along the road.

BIRCH BAY STATE PARK
Take I-5 north to the Grandview Road Exit, go west 8 miles to Jackson Road, go right to Helwig, left on Helwig to the park. Birch Bay was named in 1792 after the trees which are prolific in the park. The relatively warm water in the bay allows summertime swimming. Activities include clamming, crabbing, beach hiking, scuba diving, water skiing and shore fishing. Both Golden and Bald eagles nest in the area and are frequently seen. Park closes at dusk.

BIRCH POINT, BEACH 372
Located north of Birch Point southwest of Blaine. Access is by boat only. Upper beach is cobbles, lower beach is extensive sand flat with gravel, boulders and hardpan. Clamming. Only the tidelands are public.

BLAINE HARBOR AND BOAT LAUNCH
Take I-5 north to Blaine City Center Exit #276 (last exit in U.S.), turn left and go under freeway on Peace Portal Road. Follow signs to Blaine Harbor. Boat launch is .25 mile before the harbor. Lots of parking at Harbor Office, not at boat launch. Ample boating facilities. No beach. Access to tideflats for clamming.

COTTONWOOD BEACH COUNTY PARK (undev)
Take I-5 to Birch Bay - Lynden Road, go west to Birch Bay Drive and right 3.0 miles to Cottonwood Beach. Park is undeveloped but provides access to 1200 feet of beach and sandy tidelands.

LIGHTHOUSE MARINE COUNTY PARK
Located on the southwest tip of Point Roberts. Take Interstate 5 to Canada through the border crossing at Blaine; follow Canadian Freeway to Tsawwassen/Pt. Roberts Exit; at re-entry crossing follow signs to Lighthouse Marine Park. Features a 600 foot long boardwalk with shelters. Good view of Georgia Straight and U.S. from 30 foot tower on the boardwalk. Outdoor activities available include fishing, clamming, boating and beach combing.

MONUMENT COUNTY PARK (undev)
Located in Point Roberts at its northwest corner at the Canadian border. The park is on Marine Drive.

SEMIAHMOO COUNTY PARK
Located on Semiahmoo Sandspit near the international border crossing at Blaine. Out of Blaine take Drayton Harbor Road to the park. The park is a mixture of striking natural environment and restored historic buildings. The interpretive center, open Wednesday through Sunday, 1:00-5:00, explores both cultural and natural history. Canoe rental is available. The long sandy spit has great view of Cascades and Mt. Baker. Clam digging allowed.

SEALS

If you're walking along one of Washington's less populated beaches there is a good chance you will be spied on by a curious harbor seal. Their puppy-like faces with large human-like eyes pop above the water not far off the shoreline as they seem to study human activity on the beach. If you stop to stare back, they will slip under water where they can stay for up to 15 minutes.

Harbor seals are year round residents of Puget Sound, the Strait of Juan de Fuca, and the outer Pacific Coast. Although curious, they are shy animals and prefer quiet unpopulated areas. Seals like to "haul out" on beaches, rocks, and lografts to bask in the sun and sleep. At the slightest sign of danger they will clumsily drag their bodies back into the water where they swim with power and grace.

Because of harbor seals' speed and maneuverability, there are few prey that can escape a hungry seal. Their diet consists mostly of small fish, but also includes octopus, squid, shrimp, and crab. In captivity they eat up to fifteen pounds of seafood a day.

Swift as they are, seals sometimes fall prey to orcas (killer whales), sharks, and people. From 1947 to 1960 a bounty was placed on seals because it was believed they ate significant amounts of commercially valuable fish. During that time it is estimated 17,000 seals were killed. Today, seals are protected from killing by the federal Marine Mammal Protection Act. It is strictly unlawful to take, harass or otherwise disturb seals or any other marine mammal. Seals are now abundant in Washington waters.

A less commonly seen species of seal in Washington is the elephant seal. Male elephant seals can weigh up to 7,000 pounds while their mates average around 2,000 pounds. Aside from their huge size, male elephant seals are easily distinguished by their enormous floppy nose which hangs over their mouth—thus their name.

Elephant seals are a migrating species which feed off the coast of Washington in the spring after breeding and pupping off the coast of California. Though they are most commonly seen along the straits and around the San Juan Islands, they will occasionally visit southern Puget Sound.

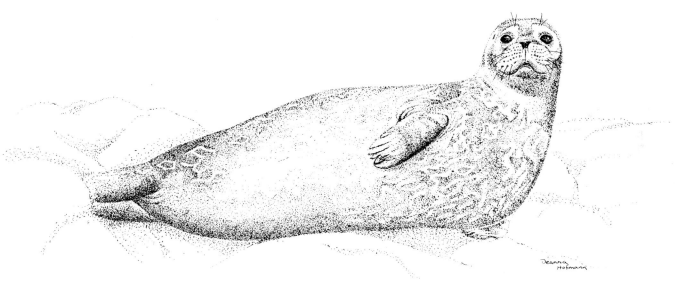

Drawing by Deanna Hofmann

WHATCOM COUNTY
SOUTH

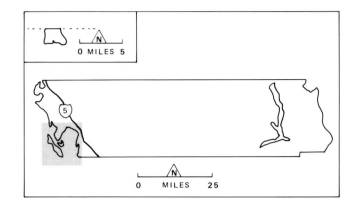

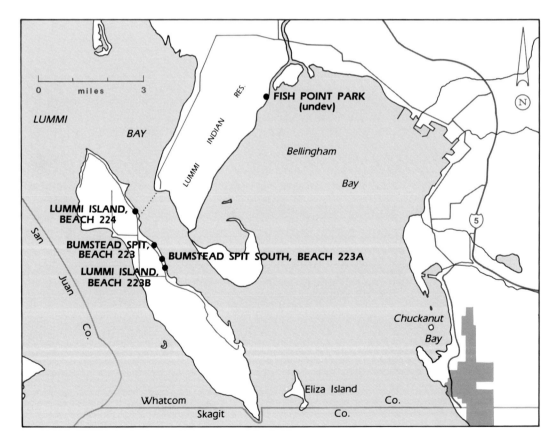

WHATCOM COUNTY
SOUTH

	ACRES	CAMP UNITS	PICNIC UNITS	RESTROOMS	FIREPITS	SWIMMING BEACH	BOAT LAUNCH (lanes)	BOAT MOORAGE (slips/buoys)	PUBLIC PIER	DRINKING WATER	VIEWPOINT	SHORELINE LENGTH (feet)	ROCK BEACH	SAND BEACH	GRAVEL BEACH	MUD BEACH	SAND DUNES	TIDE POOLS	WETLANDS	BLUFFS
BUMSTEAD SPIT SOUTH, BEACH 223A	NA											1,188	●	●						
BUMSTEAD SPIT, BEACH 223	NA											2,574	●	●	●					
FISH POINT PARK (undev)	2.0											1,750		●	●					
LUMMI ISLAND, BEACH 223B	NA											1,014		●						
LUMMI ISLAND, BEACH 224	NA											2,805	●	●	●					

BUMSTEAD SPIT SOUTH, BEACH 223A
Located on the east side of Lummi Island. Boat access only. Upper beach is pebbles and sand. Lower beach is cobbles and sand. Clamming. Only the tidelands are public.

BUMSTEAD SPIT, BEACH 223
Located on the east side of Lummi Island. Boat access only. Upper beach is cobble. Lower beach is cobble, fine sand and mud. Extensive gravel bar along much of the beach. Clamming. Only the tidelands are public.

FISH POINT PARK (undev)
Take Bakerview Road west from I-5 to Lummi Shore Road and go south 2.0 miles. Park on left. No signs. This undeveloped area has no facilities, but provides beach access on the west side of Bellingham Bay.

LUMMI ISLAND, BEACH 223B
Located on the east side of Lummi Island. Boat access only. No upland access. Beach is mostly cobbles. Clamming. Only the tidelands are public.

LUMMI ISLAND, BEACH 224
Located on the northeast shore of Lummi Island. Access is via boat. There is also access through the old ferry dock right-of-way where Whatcom county owns a 60 foot access strip south of the abandoned ferry dock. Upper beach is a series of boulders and rock outcroppings. Outer beach is a wide sand flat covered with a thick bed of eelgrass. Clamming. Except for the county right-of-way, only the tidelands are public.

WHATCOM COUNTY
SOUTH

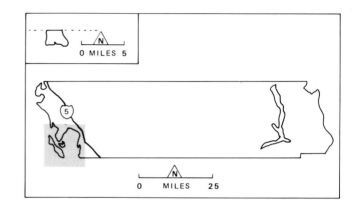

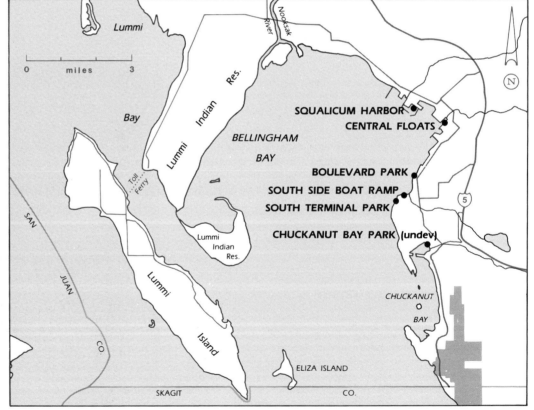

WHATCOM COUNTY
SOUTH

	ACRES	CAMP UNITS	PICNIC UNITS	RESTROOMS	FIREPITS	SWIMMING BEACH	BOAT LAUNCH (lanes)	BOAT MOORAGE (slips/buoys)	PUBLIC PIER	DRINKING WATER	VIEWPOINT	SHORELINE LENGTH (feet)	ROCK BEACH	SAND BEACH	GRAVEL BEACH	MUD BEACH	SAND DUNES	TIDE POOLS	WETLANDS	BLUFFS
BOULEVARD PARK	14.0		8	•	•			4	•	•	•	2,800		•						
CENTRAL FLOATS *Bellingham Maritime Heritage Center*	0.1							12				240			•						— more access
CHUCKANUT BAY PARK (undev)	65.0									•		9,347	•	•	•			•		•
SOUTH SIDE BOAT RAMP	2.0						2			•		150	•					•		
SOUTH TERMINAL PARK	3.0	3	•	•					•	•	•	730	•	•	•	•				
SQUALICUM HARBOR	176.0		•				4	1500	•	•	•	7,750		•						

BOULEVARD PARK
An open grassy park off State Street in south Bellingham. Nice picnic facilities. Rocky shore along Bellingham Bay. Scenic view of area available from tower in the park. Public fishing. May be shown on some maps as Bayview Marine Park.

CENTRAL FLOATS
Located at the corner of Central and Roeder Ave. in the central downtown area of Bellingham, next to the Citizen Dock. Moorage site available for visiting craft. Limited stay. Easy walking access to the city for boaters. Nearby, 0.25 mile up Whatcom Creek is the Bellingham Maritime Heritage Center which presents a comprehensive look at the Northwest's maritime industries, history and environment. It includes interpretive displays and a fish hatchery facility. *will be improved — new displays*

CHUCKANUT BAY PARK (undev)
This is a natural tideland and mudflat bay in South Bellingham off of Chuckanut Drive on Fairhaven Ave. The park, acquired in 1985, is to be left in a natural condition. Scenic view of area is available from the beach.

SOUTH SIDE BOAT RAMP
From Fairhaven City Center go west on Harris Ave., right on 6th and follow signs. Ramp is across railroad tracks. Boat launch area, owned by Port of Bellingham, has a small dock and pier. Launching fee. Parking lot is on opposite side of the tracks from the boat launch. CAUTION: watch overhead clearance (power lines) with sail boat masts.

SOUTH TERMINAL PARK
From Fairhaven City Center, follow Harris Way across railroad tracks to parking at the end of the road. Open grassy area; benches along water. Trails go south along the beach and railroad tracks. View of San Juan Islands to the west. Open until dusk. Large shelter has 2 fireplaces. No dogs allowed in the park.

SQUALICUM HARBOR
From Interstate 5 take exit 256; go west on Meridian Street, right on Squalicum Truck Route, and left on Roeden Ave. Harbor has shops, brokers, boat supplies and laundromat. First 24 hours on visitor's dock is free. Launching fee.

WHATCOM COUNTY
SOUTH

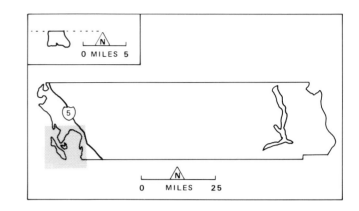

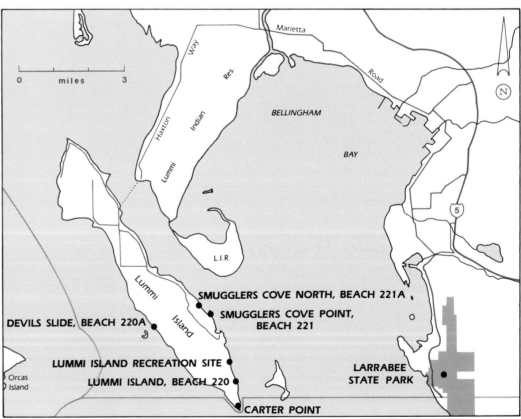

WHATCOM COUNTY
SOUTH

PUBLIC SHORE

	Acres	Camp Units	Picnic Units	Restrooms	Firepits	Swimming Beach	Boat Launch (lanes)	Boat Moorage (slips/buoys)	Public Pier	Drinking Water	Viewpoint	Shoreline Length (feet)	Rock Beach	Sand Beach	Gravel Beach	Mud Beach	Sand Dunes	Tide Pools	Wetlands	Bluffs
CARTER POINT	30.0										●	2,500								●
DEVILS SLIDE, BEACH 220A	NA											3,188	●		●					●
LARRABEE STATE PARK	1,966.4	87	40	●	●	●	2			●	●	3,600			●			●		●
LUMMI ISLAND RECREATION SITE	42.0	5		●	●	●					●	2,125			●					●
LUMMI ISLAND, BEACH 220	NA		5									23,533	●		●					
SMUGGLERS COVE NORTH, BEACH 221A	NA											4,812	●		●					
SMUGGLERS COVE POINT, BEACH 221	NA											4,481	●							●

CARTER POINT
Located at the south end of Lummi Island. Steep hillside and bluffs. Upland area is owned by the federal government as a reserve for a lighthouse. Tidelands are state owned

DEVILS SLIDE, BEACH 220A
Located on the west side of Lummi Island near Lummi Rocks. Boat access only. Narrow steep rocky beach. Uplands are very steep with several rock slides ending on the beach. There are small gravel pocket beaches on Lummi Rocks. The intertidal portion of Lummi Rocks just offshore from Beach 220A are state owned. There are submerged rocks in the area so Lummi Rocks should be approached with extreme caution. Only the tidelands are public.

LARRABEE STATE PARK
Located on scenic Highway 11 (Chuckanut Drive) 7 miles south of Bellingham. To get to boat launch go north of park to Cove Road. Rock shores have tidepools with several varieties of creatures such as starfish, crabs, barnacles, sea cucumbers, chitons, limpets and mussels. The many trails in the park include lookout views of the San Juan Islands, Mt. Baker and the North Cascades. The park has a swimming beach and nice picnic area. The park closes at dusk.

LUMMI ISLAND RECREATION SITE
Located on the S.E. shore of Lummi Island. Boating access only. Includes two nice protected cove beaches between rocky bluffs. View point from top of the bluff. A delightful spot among madronas and firs, campsites are located on the bluffs and look down on the beaches. Very scenic. See also Beach 220 which fronts this area.

LUMMI ISLAND, BEACH 220
Located on the south end of Lummi Island. Encompasses the Lummi Island Recreation Site. Mostly rocky with some coves of gravel. Except for the recreation site only the tidelands are public.

SMUGGLERS COVE NORTH, BEACH 221A
Located on the east side of Lummi Island. Boat access only. Mostly rocky beach with two gravel pocket beaches. Only the tidelands are public.

SMUGGLERS COVE POINT, BEACH 221
Located on the point east of Smugglers Cove on Lummi Island. Boat access only. Generally steep rocky cliff with a few pocket beaches. Only the tidelands are public.

RED TIDES

Red tides in Washington coastal waters are the result of blooms of microscopic single-celled plants called dinoflagellates. At times, dinoflagellates become so abundant that they actually give the water a reddish tinge. One of the dinoflagellates, *Gonyaulax catenella*, contains a powerful neurotoxin, called saxitoxin, which is responsible for paralytic shellfish poisoning (PSP) and the closure of shellfish beds. Saxitoxin is a thousand times more potent than cyanide and fifty times more toxic than strychnine.

Although humans do not consume *Gonyaulax* directly, clams, oysters, mussels and other filter feeding animals feed on *Gonyaulax* and concentrate it in their tissues. If a person eats the contaminated shellfish, a tiny amount of the toxin is enough to make him or her quite sick and slightly larger amounts can be fatal.

Part of the danger of red tide is the difficulty detecting it. Red tide can be present when the water looks clear and clams appear healthy. Not only are there many things that can turn the water a reddish tinge, red tide does not always turn the water red. Toxins in shellfish can only be determined by sending shellfish samples to the State Public Shellfish Laboratory where they are analyzed by a bioassay method.

PSP affects the nervous system and symptoms can begin within minutes after eating contaminated shellfish. Initially there will be a numbness in the lips and tongue and tingling sensations in the fingers and extremities. Later, symptoms will develop into breathing difficulties, loss of muscular control, and nausea. If these symptoms occur, immediately induce vomiting, take a quick-acting laxative and go to the nearest emergency room. There is no antidote for PSP, but if the patient survives a critical twelve-hour period there will probably be no lasting complications. All shellfish poisonings should be reported to the State Department of Social and Health Services to protect others from getting poisoned.

Paralytic shellfish poisoning in humans has been recorded throughout the world for centuries. Coastal Indians had taboos and legends associated with eating shellfish at certain places and at certain times of the year. There is a legend that Indians disposed of a large group of troublesome Russian explorers by inviting them to a feast of contaminated shellfish.

Today we know little more than the Indians did about what triggers a bloom of *Gonyaulax*. Many researchers feel it is a combination of factors, among them temperature, salinity, and nutrients. Blooms are most common during late spring, summer, and early fall, but they can also occur in the winter.

As of summer 1982, the only waters that had not been periodically closed to shellfish harvesting because of contamination are Hood Canal and Puget Sound south of the Tacoma Narrows bridge. Nevertheless, evidence shows that red tide is definittely creeping south.

The Washington Department of Social and Health Services monitors red tide and the closure of beaches where shellfish are harvested. Commercial harvesters are required to submit samples of shellfish every two weeks, and more frequently when toxin levels are high. It is impossible to test shellfish on popular public harvesting beaches frequently because the department doesn't have the manpower. In many cases it must rely on volunteers to do the sample collecting. Because of this, the department closes beaches before levels are toxic and your favorite beach may stay closed for quite a while after danger of red tide has disappeared. If you have questions about a particular beach it's a good idea to call before you dig. The Department of Social and Health Services twenty-four hour toll-free red tide hotline is: 1-800-562-5632

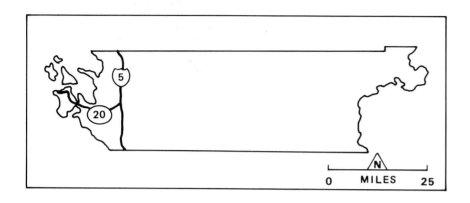

SKAGIT COUNTY

The total length of saltwater shoreline within Skagit County is 127 miles, which includes several of the San Juan Islands that lie within the county's boundary. The Skagit River Delta and Padilla Bay, both extensive areas of estuary, wetland, and winding channels, represent some of the county's most impressive coastal features.

Padilla Bay is one of seventeen areas in the country designated as a National Estuarine Sanctuary. The bay and the interpretive center are excellent places to explore and learn about the coastal zone. (See special write-up on Padilla Bay.)

Just south of Padilla Bay, through Swinomish Channel, is the Skagit Habitat Management Area (HMA). The Skagit River, which feeds the HMA at the delta, is second only to the Columbia in hydraulic volume and the number of salmon it produces in Washington. Inland, the Skagit provides an important winter habitat for bald eagles. Eagles are also seen at the Skagit delta and along the county's coastal zone year-round. Thousands of waterfowl make their home at the Habitat Management Area or use it as a stopover on their annual migration. Some of the more exotic birds found around the Skagit are two species of swan and the Siberian snow goose that winters there.

Two coastal communities in Skagit County that deserve mention are La Conner and Anacortes. La Conner is a historic and picturesque town that is known for its many antique shops and galleries. Along the LaConner waterfront on Swinomish Channel are several sites for picnics.

Nearby, Anacortes serves as the ferry terminal for the San Juan Islands and British Columbia. Anacortes has several large parks, a huge marina, and historic points of interest. The town is well equipped for visitors: there are several motels and interesting shops in which to browse. Road signs lead visitors to the ferry terminal and to shoreline public access areas.

The two major islands of Skagit County, Guemes and Cypress, both have waterfront parks. Cypress Island, named by Captain George Vancouver when he mistook the juniper trees there for cypress, has been largely left in a pristine state and is only accessible by private boat. Guemes Island has a small residential community and is accessible by the privately owned Guemes Island Ferry service.

SKAGIT COUNTY
NORTH

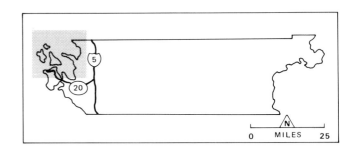

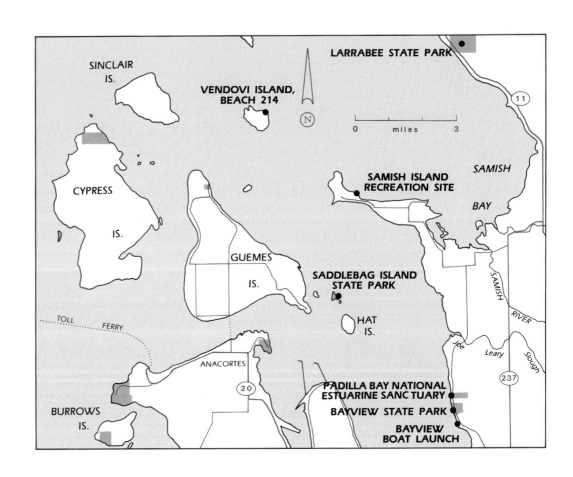

SKAGIT COUNTY
NORTH

	ACRES	CAMP UNITS	PICNIC UNITS	RESTROOMS	FIREPITS	SWIMMING BEACH	BOAT LAUNCH (lanes)	BOAT MOORAGE (slips/buoys)	PUBLIC PIER	DRINKING WATER	VIEWPOINT	SHORELINE LENGTH (feet)	ROCK BEACH	SAND BEACH	GRAVEL BEACH	MUD BEACH	SAND DUNES	TIDE POOLS	WETLANDS	BLUFFS
BAYVIEW BOAT LAUNCH	0.3						1					20	•							
BAYVIEW STATE PARK	23.9	110	59	•	•	•				•		3,640	•	•	•	•				
LARRABEE STATE PARK	NA											NA								
PADILLA BAY NATIONAL ESTUARINE SANCTUARY	11,000.0		•				1		•	•		NA				•			•	
SADDLEBAG ISLAND STATE PARK	23.2	5		•	•	•				•		6,250	•		•					
SAMISH ISLAND RECREATION SITE	0.5		•									1,436	•		•					
VENDOVI ISLAND, BEACH 214	NA											12,550	•		•					

BAYVIEW BOAT LAUNCH
Located 0.5 mile south of Bayview State Park on the Bayview/Edison Road. The site has a concrete boat launch, rocky beach and limited parking. No facilities other than the boat ramp which is in disrepair, and not very usable.

BAYVIEW STATE PARK
Take I-5 to Chuckanut Drive exit; take Wilson Road west 6 miles to Bayview-Edison Drive; turn right (north) and go 0.5 miles to the park. Camping is on opposite side of road from beach. There are open grassy fields at the camping area and near the beach. Large play area for children. Swimming in cold saltwater. Saltwater fishing.

LARRABEE STATE PARK
A heavily wooded park straddling the Skagit and Whatcom County line on Chuckanut Drive. Hiking trails through the woods to beaches and tidepools. This is a good site to view marine life. There is a large grassy and wooded picnic area with two covered shelters. Group camping sites can be reserved. Excellent place to view Samish Bay and the San Juan Islands. Also see the Larrabee State Park write-up under Whatcom County.

PADILLA BAY NATIONAL ESTUARINE SANCTUARY
Located at Padilla Bay; access is off Highway 20. Take Exit 230 from I-5 and drive 5.5 miles west; watch for Padilla Bay N.E.S. sign at turnoff to the right. Breazeale Interpretive Center provides opportunity for education, research and "hands on" learning about marine ecosystems. Wildlife habitat areas and nature trails. Hours are Wed. - Sun., 10:00 - 5:00, all year. Also see write-up: "Padilla Bay."

SADDLEBAG ISLAND STATE PARK
Boating access only. A rocky island with fir trees. Two crescent shaped beaches. Popular as a crabbing spot. The best access is to the north side.

SAMISH ISLAND RECREATION SITE
Turn right off Old Samish Point Road to Samish Island Road; go 0.5 mile to dead end; beach is on the right. Wooded bluff overlooking Samish Bay. Clam digging and crabbing. No camping. Limited parking and turn-around space. Short, steep trail down the bluff to a gravel beach.

VENDOVI ISLAND, BEACH 214
Located two miles northeast of Guemes Island. Access to the state owned tidelands is by boat only. Beach is mostly rocky with tide pools. There are three small gravel beaches with opportunities for clam digging.

PADILLA BAY

In recognition of its environmental value as a pristine estuary, Padilla Bay is established as one of seventeen National Estuarine Sanctuaries in the country. The sanctuary operates under shared state and federal funding and is managed by the Washington Department of Ecology. The proposed boundary encloses 11,000 acres of tideland, most of which are relatively undisturbed mudflats. The estuary is a nursery, migration stop, and home for many birds, fishes and mammals.

Ten habitat types with characteristic plant and animal communities have been identified at Padilla Bay. They include open marine waters, subtidal sand and mud, eelgrass beds, exposed mudflats, salt marshes, beaches, rocky shores, and forested and non-forested uplands.

As many as 57 species of fish, 14 species of mammals, and 241 species of birds have been spotted at Padilla Bay. Its strategic location on the Pacific Flyway and shallow protected bay which is covered with eelgrass beds make it an important rest stop for migrating black brant and many other waterfowl. An average of 80,000 ducks spend the winter there.

Edna Breazeale, a retired school teacher, donated a 64-acre site with buildings that made development of the Breazeale-Padilla Bay Interpretive Center possible. The center houses a resource library containing books on ecology, natural history, marine life and birds, as well as publications associated with the bay and Puget Sound. There is also a theater, laboratory, exhibit area and "hands-on" room where visitors can touch and look closely at features of the natural community. There is a mile long upland trail and one access site to the bay. More trail and access development is proposed.

Education is an important aspect of the sanctuary. Its purpose is to encourage people of all backgrounds to learn about the natural resources and ecology of the coastal zone. The center offers a "Mini Explorers" program for kids 3-5; a "Junior Ecologists" program for kids 6-11; educational workshops for the general public, including summer family workshops called "Estuary Exploration;" extensive school programs for all grades, and a weekly film series. It is hoped that these programs will foster a better appreciation and understanding of the resources and ecological processes of estuaries, and in turn lead to sound decision making and effective management of our vital shorelands.

For more information about Padilla Bay activities contact:

Padilla Bay National Estuarine Sanctuary
1043 Bay View-Edison Road
Mt. Vernon, WA 98273
Phone: (206)428-1558

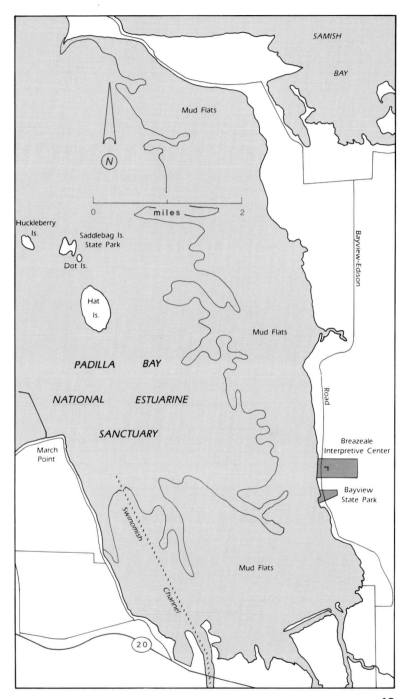

SKAGIT COUNTY
NORTH

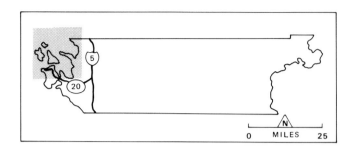

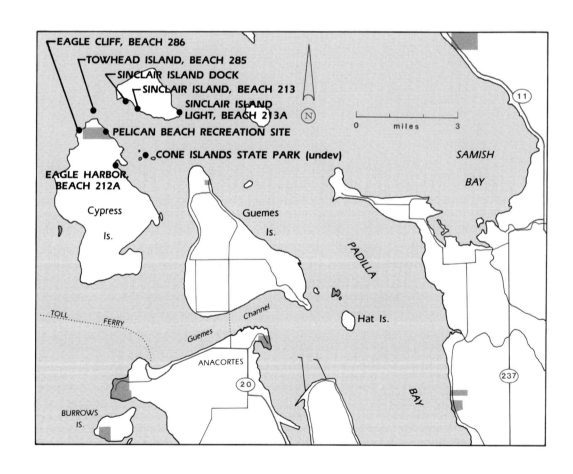

SKAGIT COUNTY
NORTH

	ACRES	CAMP UNITS	PICNIC UNITS	RESTROOMS	FIREPITS	SWIMMING BEACH	BOAT LAUNCH (lanes)	BOAT MOORAGE (slips/buoys)	PUBLIC PIER	DRINKING WATER	VIEWPOINT	SHORELINE LENGTH (feet)	ROCK BEACH	SAND BEACH	GRAVEL BEACH	MUD BEACH	SAND DUNES	TIDE POOLS	WETLANDS	BLUFFS
CONE ISLANDS STATE PARK (undev)	9.9							2				2,500	•		•					
EAGLE CLIFF, BEACH 286	NA											9,220	•		•					•
EAGLE HARBOR, BEACH 212A	NA											2,118			•	•				
PELICAN BEACH RECREATION SITE	211.0	4	3	•	•			3				7,207			•					
SINCLAIR ISLAND DOCK	NA							2				NA		•						
SINCLAIR ISLAND LIGHT, BEACH 213A	NA											2,831			•					
SINCLAIR ISLAND, BEACH 213	NA											5,200	•		•				•	
TOWHEAD ISLAND, BEACH 285	NA											1,300			•					

CONE ISLANDS STATE PARK (undev)
Boat access only which is limited because of the rocky bluff shoreline. The park comprises all of Cone Islands, which are four small islands with mostly rock ledge beaches. A small gravel beach provides an access point. CAUTION: difficult access; not a recommended site.

EAGLE CLIFF, BEACH 286
Located on the northwest corner of Cypress Island. Access is by boat only. The uplands around Eagle Cliff are state owned and part of the Pelican Beach Recreation Site. The beach is a series of rock cliffs interspersed with nice pea gravel and rock beaches. CAUTION: there is no trail up Eagle Cliff; use extreme care in climbing the cliff.

EAGLE HARBOR, BEACH 212A
Located on the eastern side of Cypress Island. Boat access only. Beach is typically narrow and steep consisting mostly of mud, gravel and cobbles. Only the tidelands are public.

PELICAN BEACH RECREATION SITE
Located on the N.E. side of Cypress Island. Access is by boat only. The beach is a combination of rock cliffs and large pea gravel. Clam digging. Scenic vistas of mountains and various islands.

SINCLAIR ISLAND DOCK
Located on the southwest corner of Sinclair Island. This site is an accretion beach which is well suited for recreational uses. There are excellent views across Rosario Strait to the San Juan Islands. Beach is good for walking and beachcombing. The county owned dock may be used for loading and unloading but not for extended moorage.

SINCLAIR ISLAND LIGHT, BEACH 213A
Located on the east side of Sinclair Island. There is upland access from State Department of Game lands on the northeast end of the beach. Unguided nature walk. Beaches are gravel with a rock outcrop in the center.

SINCLAIR ISLAND, BEACH 213
Located on the southern shore of Sinclair Island. Access is by boat only. A series of pocket beaches interspersed with rock outcrops and headlands. Only the tidelands are public.

TOWHEAD ISLAND, BEACH 285
Located just off the north tip of Cypress Island. Boat Access only. The uplands are private, but the tidelands are public.

SKAGIT COUNTY
NORTH

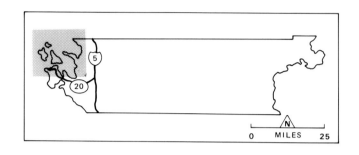

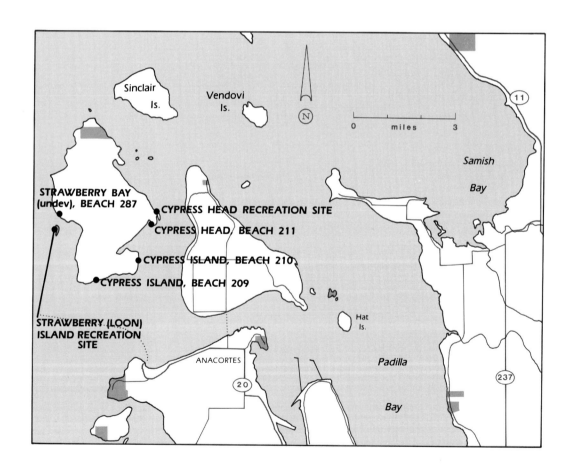

SKAGIT COUNTY
NORTH

	ACRES	CAMP UNITS	PICNIC UNITS	RESTROOMS	FIREPITS	SWIMMING BEACH	BOAT LAUNCH (lanes)	BOAT MOORAGE (slips/buoys)	PUBLIC PIER	DRINKING WATER	VIEWPOINT	SHORELINE LENGTH (feet)	ROCK BEACH	SAND BEACH	GRAVEL BEACH	MUD BEACH	SAND DUNES	TIDE POOLS	WETLANDS	BLUFFS
CYPRESS HEAD RECREATION SITE	16.5	5	7	●				4				4,780		●				●		●
CYPRESS HEAD, BEACH 211	NA											15,652	●		●					●
CYPRESS ISLAND, BEACH 209	NA											1,635	●	●						
CYPRESS ISLAND, BEACH 210	NA											5,320	●							●
STRAWBERRY (LOON) ISLAND RECREATION SITE	8.8		●									4,290	●		●					
STRAWBERRY BAY (undev), BEACH 287	NA											8,872			●					

CYPRESS HEAD RECREATION SITE
Located at the eastern most point of Cypress Island. Boating access only. A rocky promontory connected to the main island by a gravel isthmus with gravely crescent beaches on both sides. The entire shoreline of the head is public, but there is no access to the main island. The rocky shore has tidepools which are hard to get to. Mooring float and dock. Firs and madronas on the point. A very clean and well maintained site.

CYPRESS HEAD, BEACH 211
Located at Cypress Head on Cypress Island. Includes the Cypress Head Recreation Site. Access is by boat only. The beach is typically steep and composed of angular rubble. There are some sections of gravel. The south portion of this beach includes North Deepwater Bay, an excellent area for bird watching and nature study. There is a pocket beach which can be used for picnicking.

CYPRESS ISLAND, BEACH 209
Located on the south side of Cypress Island. Boat access only. Beach is sand and gravel with kelp beds off shore. Only the tidelands are public.

CYPRESS ISLAND, BEACH 210
Located on the southeast shore of Cypress Island. Boat access only. Mostly a rock cliff with a gravel beach. A big kelp bed off shore. Only the tidelands are public.

STRAWBERRY (LOON) ISLAND RECREATION SITE
Located 0.5 mile west of Cypress Island. Boating access only. Primitive camping. Fairweather moorage is available at the south end. The setting is outstanding with excellent views across Rosario Strait to the San Juan Islands. Most of the island has a steep and rocky shore. The boat landing area is a small pocket beach. CAUTION: submerged rocks and strong tidal currents.

STRAWBERRY BAY (undev), BEACH 287
Located on the southwest side of Cypress Island north of Strawberry Bay. Access is by boat only. Nice beach. Views of Rosario Strait, and Strawberry Island. Opportunities for nature study and bird watching. Clam digging.

53

SKAGIT COUNTY
NORTH

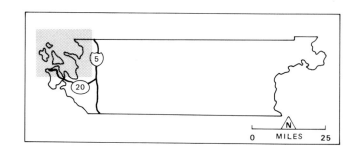

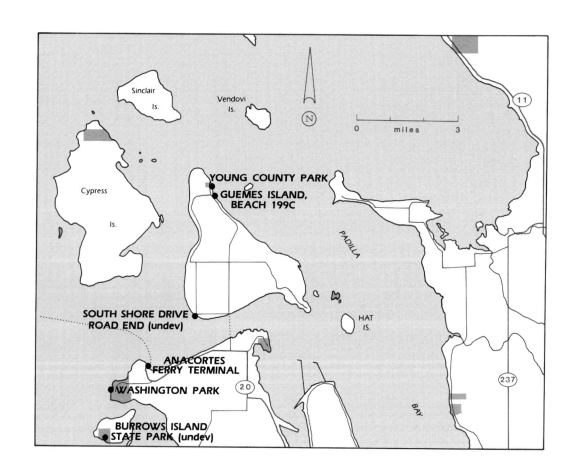

SKAGIT COUNTY
NORTH

	ACRES	CAMP UNITS	PICNIC UNITS	RESTROOMS	FIREPITS	SWIMMING BEACH	BOAT LAUNCH (lanes)	BOAT MOORAGE (slips/buoys)	PUBLIC PIER	DRINKING WATER	VIEWPOINT	SHORELINE LENGTH (feet)	ROCK BEACH	SAND BEACH	GRAVEL BEACH	MUD BEACH	SAND DUNES	TIDE POOLS	WETLANDS	BLUFFS
ANACORTES FERRY TERMINAL	NA		•								•	NA		•				•	•	
BURROWS ISLAND STATE PARK (undev)	40.5											1,000	•							•
GUEMES ISLAND, BEACH 199C	NA											1,736		•	•					
SOUTH SHORE DRIVE ROAD END (undev)	NA											NA		•	•					
WASHINGTON PARK	220.0	79	38	•	•	•	2	4		•	•	40,560	•	•	•		•			
YOUNG COUNTY PARK	11.0		4	•	•							500		•						

ANACORTES FERRY TERMINAL
Located at the terminal for ferries to the San Juan Islands and Vancouver Island, Canada, this facility is primarily intended for passengers, but it is possible to park here and walk down to a wetland and beach in Ship Harbor which is immediately east of the ferry terminal. The area is slated for a major marina development, but will contain extensive public access when finished.

BURROWS ISLAND STATE PARK (undev)
No safe boat access is presently available. No shoreline or upland recreational facilities developed. CAUTION: very dangerous boating conditions during tidal changes and windy days.

GUEMES ISLAND, BEACH 199C
Located on the northeast side of Guemes Island. Upland access is via Young County Park. Beach is gravel and sand with eel grass and kelp off shore. Only the tidelands are public.

SOUTH SHORE DRIVE ROAD END (undev)
Located on Guemes Island which is accessible by ferry from Anacortes. This site provides access to over one mile of beach on the west side of Guemes Island. No developed facilities. The beach is good for walking. Scenic view of Rosario Strait.

WASHINGTON PARK
Take Highway 20 to Washington Park, which is located just southwest of the Anacortes Ferry Terminal. Large park in old-growth forest. Hiking trail and playground equipment. Major boat launch for access to San Juan Islands. The park gets very crowded on summer weekends.

YOUNG COUNTY PARK
Located at the north end of the Guemes Island Road on Guemes Island. Park has a small picnic area and a grass playfield. Opportunity for shoreline fishing. Excellent scenic views to the east and north. This park also provides access to Beach 199C.

SKAGIT COUNTY
NORTH

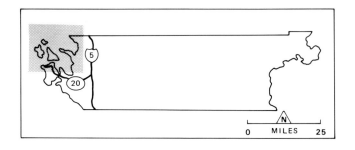

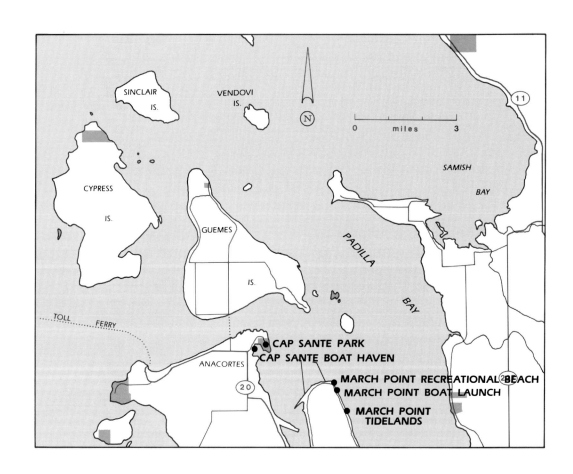

SKAGIT COUNTY
NORTH

	ACRES	CAMP UNITS	PICNIC UNITS	RESTROOMS	FIREPITS	SWIMMING BEACH	BOAT LAUNCH (lanes)	BOAT MOORAGE (slips/buoys)	PUBLIC PIER	DRINKING WATER	VIEWPOINT	SHORELINE LENGTH (feet)	ROCK BEACH	SAND BEACH	GRAVEL BEACH	MUD BEACH	SAND DUNES	TIDE POOLS	WETLANDS	BLUFFS
CAP SANTE BOAT HAVEN	60.0		●					926	●			3,500		●	●					
CAP SANTE PARK	40.0	4									●	15,000	●		●	●				
MARCH POINT RECREATIONAL BEACH	NA			●			1					5,280			●					
MARCH POINT TIDELANDS	NA											NA		●						

CAP SANTE BOAT HAVEN
Take Highway 20 west to Commercial Ave., into Anacortes; turn right on 11th and follow to park. Full marina facilities. Shops, restaurants and boating supplies are available.

CAP SANTE PARK
Take Commercial Ave. off of Highway 20 to 4th Street and follow signs to the viewpoint. To get to the beach, take Highway 20 to Commercial Ave., to 4th Street, and go right on T Ave., veer left at Cap Sante Marina sign and go down a gravel road. Park at designated parking area. An unmarked and undeveloped park. Trail to viewpoint, where there are good views of Anacortes, the San Juan Islands and Puget Sound. *Funded to provide better access along 1500 ft. of shoreline*

MARCH POINT RECREATIONAL BEACH
Includes the beach front and associated tidelands on the northern tip of March Point. Private, but self-contained R.V. camping and beach use is allowed. View of Mt. Baker, Cascades and Padilla Bay. Popular for fishing and shellfish gathering. The area is bordered on the west by tanker docking facilities and on the east by a primitive Department of Game boat launch ramp. The ramp is suitable for small boats, trucks and four-wheel drive vehicles. This is a popular recreation site but may be withdrawn from public use without notice. Visitors should use the area with care and thus assure its continued availability.

MARCH POINT TIDELANDS
About two miles of tidelands which are used by the public. Public use is limited because of private upland ownership. People get to the beach at the north end, about 0.5 miles north of Kavanaugh Road where there is some off-road parking and little chance of conflict with upland property owners. Excellent views of the North Cascades and the Canadian Coast Range.

SKAGIT COUNTY
SOUTH

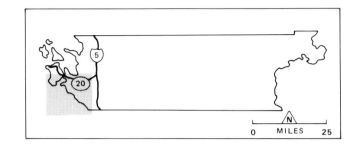

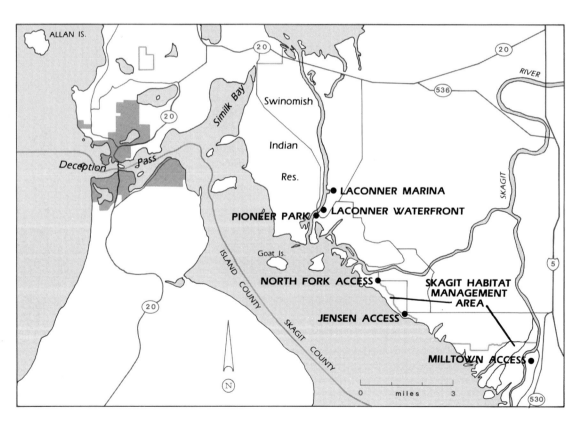

SKAGIT COUNTY
SOUTH

PUBLIC SHORE

	ACRES	CAMP UNITS	PICNIC UNITS	RESTROOMS	FIREPITS	SWIMMING BEACH	BOAT LAUNCH (lanes)	BOAT MOORAGE (slips/buoys)	PUBLIC PIER	DRINKING WATER	VIEWPOINT	SHORELINE LENGTH (feet)	ROCK BEACH	SAND BEACH	GRAVEL BEACH	MUD BEACH	SAND DUNES	TIDE POOLS	WETLANDS	BLUFFS
JENSEN ACCESS	NA											NA			•				•	
LACONNER MARINA	66.0		2	•	•			572	•	•	•	5,320								
LACONNER WATERFRONT	0.3		3									NA								
MILLTOWN ACCESS	NA											NA			•				•	
NORTH FORK ACCESS	NA											NA			•				•	
PIONEER PARK	NA	10	25	•	•		1					NA	•	•						
SKAGIT HABITAT MANAGEMENT AREA	10,160.0			•			1					NA			•				•	

JENSEN ACCESS
Located off of the Maupin Road, this undeveloped access point is part of the Skagit Wildlife Habitat Area.

LACONNER MARINA
Located at the north end of LaConner. Turn north on 3rd Ave. Long term leased moorage and a public pier for transient travelers. Full marina services. It is possible to walk into town and explore the historic buildings.

LACONNER WATERFRONT
There are 3 small public areas in Laconner on the waterfront. Each has picnic tables and some have piers. A great place to explore on foot—small shops and old buildings. An excellent place to fish in the Swinomish Channel.

MILLTOWN ACCESS
Located 0.25 mile north of Milltown off Old Highway 99. Provides an access point to the Skagit Wildlife Habitat Area.

NORTH FORK ACCESS
Located at the northern end of the Skagit Wildlife Habitat Area, the site offers access to 13,000 acres of intertidal estuarine mud flats. There is no trail system at this location so users must be prepared for muddy conditions if they wish to visit the flats. Excellent bird watching area. The best way to explore the area is by small portable boat.

PIONEER PARK
In Laconner; take Caledonia Road to 3rd and take left at "T" in the road; turn right to the boat launch and left to the park. Limited camping facilities on the hill by the bridge. The boat launch is on the Swinomish Channel under the bridge.

SKAGIT HABITAT MANAGEMENT AREA
From I-5 take Exit exit 221; travel west through Conway on Fir Island Road; turn left on Mann Road and go south 1 mile to the headquarters where information habit area access is available. There is a pavilion for classes, trails to wetlands and a boat launch into the Skagit River. The area is open for hunting in season. Bird watching is excellent. During late winter this is an excellent place for observing migratory waterfowl, notably, snow geese. There are several accesses to the area which are listed separately. A portion of the area is in Snohomish County. A State Department of Game Conservation License is required to use this area.

ESTUARIES

To the casual observer, there is not much to catch your eye: relatively featureless grassy shores when the tide is high, and expanses of flat black mud when the tide is low. With a closer look, however, one finds the estuary teeming with life. In fact, acre for acre, estuaries and their associated wetlands are among the most biologically productive ecosystems on earth.

Estuaries are transition zones between land and sea. They are found in sheltered bays, inlets, and lagoons where freshwater rivers and streams meet and mix with the salt water, forming a melting pot of organic and mineral nutrients. The nutrient-rich soup of the estuary enhances the growth of plankton and plants which, in turn, nourish an abundance of other estuarine organisims including oysters, clam, crab, salmon, and waterfowl. These species may then sustain marine birds and mammals, animals along the shore, and humans.

Estuarine Habitats

The most interesting way to explore an estuary is to view it in different biological zones. Each zone is host to dominant plant and animal species whose distribution is determined by salinity (the amount of salt present in the water and soil of an estuary), elevation and tidal inundation (which determine the amount of time an area is submerged or exposed by the tides), and substrate (in particular the size of the soil particles). These characteristics, particularly salinity and inundation, may change with seasonal fluctuations in rainfall and tidal cycles.

We will explore biological zones as though they are separate and distinct. In reality, though, they blend together to form one large, interconnected system without distinct boundaries. The diagram will help you visualize how these zones interrelate with one another. The following descriptions progress from deep water habitats to dry land.

The subtidal, or deep water, area of an estuary usually has a substrate of mud, sand, or gravel. Pacific oysters thrive where the estuary floor is hard mud and gravel. They lie on the bottom and graze on green plankton and other nutrients that are delivered daily by the tide and streams. Many of the estuaries in Washington State produce oysters. Willapa Bay in Pacific County is the largest oyster producing area on the entire west coast of the United States.

If the estuary floor is predominately sand, it will host sand dollars, snails, and several species of clams. A muddier floor has more organic matter and can support large populations and a wide diversity of invertebrates (clams, shrimps, and worms), many of which live in no other environment.

The shallow waters of the subtidal zone are used by seals, waterfowl, and many species of game fish for feeding, breeding, and maturing. This area is of particular significance to salmon. Young salmon travel down the streams in which they were born to estuaries where the shallow waters serve as a refuge from larger predators. A young salmon spending several weeks in nutritious estuarine waters will often triple in size. The estuary environment also allows the salmon to make a gradual adjustment from fresh to salt water.

The salmon often find food and cover in eelgrass beds which grow in the shallow subtidal and lower intertidal regions. Eelgrass is not restricted to the estuarine environment, but it grows best in quiet protected harbors with sand and mud bottoms. The roots of dense eelgrass beds form a thick mat and hold sediments which host highly productive bacteria. These bacteria are an important source of nutrition for many small invertebrates. The result is a high density and diversity of invertebrates, many of which go unnoticed because they are microscopic. The eelgrass bed is a haven for small fish and crab who feed on the invertebrates and find cover in the grasses.

The upper limits of eelgrass beds often mark the beginning of the intertidal region: the area between the low tide line and the high tide line. This area is usually a broad sloping region where bare mudflat gradually rises to salt marsh vegetation which forms the estuary's wetlands. Tidal inundation varies with elevation but is usually enough so that salt marsh communities are dominated by halophytes—plants that are tolerant of saline environments.

The lowest elevations of the intertidal region are sparsely colonized by pickleweed, sand spurry, and arrowgrass. These

succulent-leaved plants which are well adapted to long periods of submergence when the tide is high and dessication when the tide is low are the typical "pioneers" of the intertidal area.

Shorebirds and herons, along with an occasional raccoon, are often seen here roving the beach for dinner.

Higher up the beach, gumweed, jaumea, stellaria, saltwort, goose tongue and salt grass intermingle with the other pioneer species to form low, dense mats dissected by shallow drainage channels. As the elevation rises, sedges move in and rushes begin to take root.

Beyond the reach of regular tidal inundation, layers of vegetation build up forming peat soil topped by a rich humus layer. Drainage channels in this area may reach depths of three to five feet. High salinity here is moderated by precipitation and freshwater runoff from upland drainage. Succulent plants are replaced by less salt-tolerant species such as tufted hairgrass, red fescue, alkali grass, pacific silverweed, and marsh aster.

The highest, or least saline areas may be colonized by cattails and bulrushes.

The entire estuarine environment—shallow waters, mudflats, and salt marshes—serves as a major resting and refueling station for thousands of ducks, geese and shorebirds. Unfortunately, many of America's estuaries have been lost to indiscriminate development. Washington has two major coastal estuaries that still have large relatively natural areas remaining: Willapa Bay and Grays Harbor. Both are on the Pacific flyway, the main migration route for west coast waterfowl. Pintails, brants, teals, mallards, Canada geese, and trumpeter swans are a few of the species that stop here to graze on seeds, grasses, and small invertebrates.

Migrating visitors usually arrive on their northward journey around April or May, and on their southward return in September or October. Other birds, such as the teals and widgeons, use the edges of Washington's estuaries as a place to build nests and raise their young.

Conflicting Uses of Estuaries

The same qualities that make estuaries a haven for young salmon, oysters, and migrating waterfowl make it a haven for boaters, developers, and industrialists: protected harbors, a supply of fresh water, large expanses of flat unforested land, and direct access to the ocean. These conditions have created competition between humans and nature that has been strongly influenced by the mistaken belief that estuaries, in their natural state, are worthless. As a result, estuaries have been losing the battle against human development.

Ever since settlers arrived in Washington they have used several techniques to "reclaim" estuarine land and make it "useful." In many areas dikes have been built. Dikes are walls, usually made out of earth or rock, built to hold tidal salt water out of a wetland area. The rich marsh soil can then be used for farming. This practice was particularly prevalent in Skagit County. Draining, another common practice, usually accompanies diking. Ditches are built to entrapped fresh water and to prevent a freshwater wetland from forming where formerly there was a saltwater wetland.

Because the open water of an estuary usually is not deep enough for boats, and the wetlands are not high and dry enough for upland development, they are often dredged and filled to make them suitable for other uses. Estuary floors are dredged, or dug out, to deepen them for use as ports. The mud and silt "dredge spoils" are commonly used to fill adjacent wetlands which can then be developed. Many of the towns along the shore of Puget Sound are built on "fill."

It is estimated that over 50 percent of the original estuaries in the United States have been lost to these "reclamation" activities. On the outer coast of Washington, about 36 percent of Willapa Bay's wetlands have been lost to diking, and over 30 percent of the marshes in Grays Harbor have been lost to dredging and filling. In eleven major Puget Sound estuaries, only 52 percent of the original salt marsh and mudflat habitat remains; in three of the eleven areas, losses of salt marsh habitat is close to 100 percent.

Washington is fortunate to have a few significant estuaries that are now protected as refuges; Willapa Bay, Skagit Bay, Nisqually Delta, and Padilla Bay. Although all of these areas have suffered some disturbance and loss of habitat, much of their natural character has been preserved. Visitors to these areas can view the many different plant and animal species that make their home in the estuarine environment.

Recently, new light has been shed on the traditional view of estuaries and wetlands as "wastelands." Research has shown that not only do estuaries provide habitat for plants and animals, but also provide valuable "services" such as flood control, water purification, absorption of storm energy, and erosion control. As our understanding of these systems in their natural state has grown, so has our willingness to protect and preserve them.

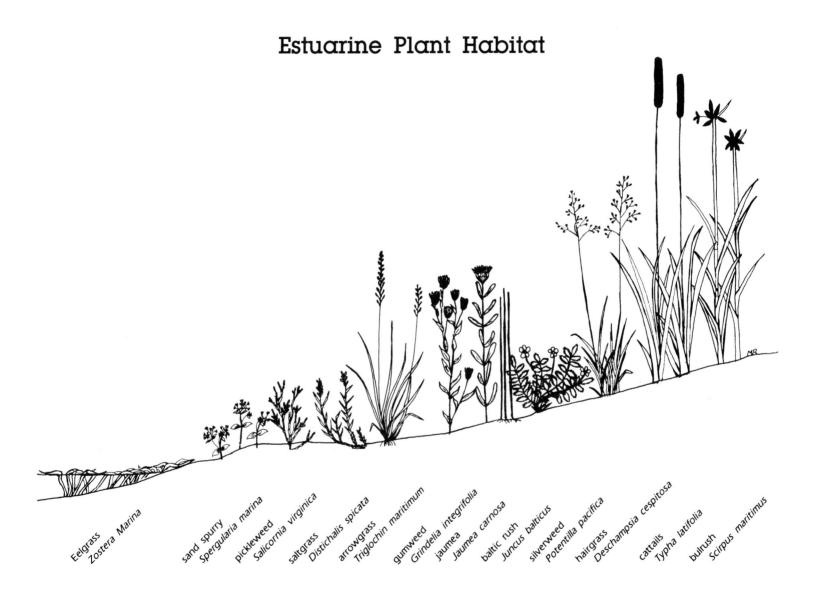

SKAGIT COUNTY
SOUTH

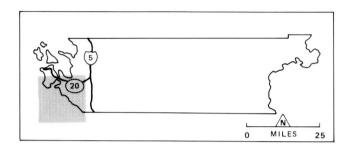

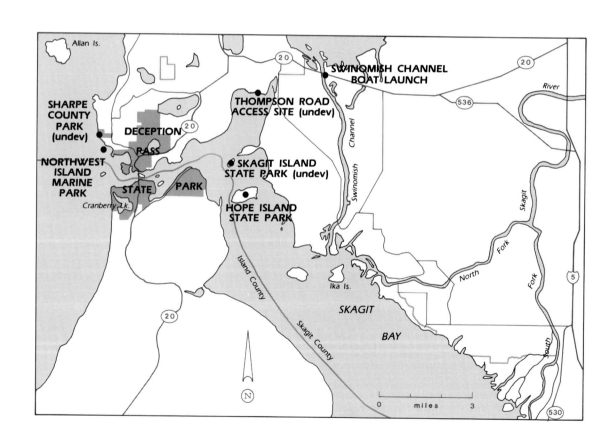

SKAGIT COUNTY
SOUTH

PUBLIC SHORE

	ACRES	CAMP UNITS	PICNIC UNITS	RESTROOMS	FIREPITS	SWIMMING BEACH	BOAT LAUNCH (lanes)	BOAT MOORAGE (slips/buoys)	PUBLIC PIER	DRINKING WATER	VIEWPOINT	SHORELINE LENGTH (feet)	ROCK BEACH	SAND BEACH	GRAVEL BEACH	MUD BEACH	SAND DUNES	TIDE POOLS	WETLANDS	BLUFFS
DECEPTION PASS STATE PARK	NA											NA								
HOPE ISLAND STATE PARK	40.5	5	•					4				100	•							
NORTHWEST ISLAND MARINE PARK	1.0											700	•	•						•
SHARPE COUNTY PARK (undev)	74.8		2	•							•	2,640	•						•	•
SKAGIT ISLAND STATE PARK (undev)	40.5											100	•		•					
SWINOMISH CHANNEL BOAT LAUNCH	NA		2	•			1					NA	•	•						
THOMPSON ROAD ACCESS SITE (undev)	NA											NA	•		•					

DECEPTION PASS STATE PARK
The northern half of this large park is located in Skagit County. The southern half is in Island County. The Rosario Beach area offers complete picnicking facilities and is suitable for walking, studying tidepool life and saltwater fishing. Park has several interesting historic exhibits and a freshwater lake which provides swimming. Also see the write-up under Island County.

HOPE ISLAND STATE PARK
Located in Skagit Bay 0.5 mile off Snee-oosh Point midway between the mainland and Whidbey Island. Access by boat only. Several small beaches on the north side of the island, otherwise rocky.

NORTHWEST ISLAND MARINE PARK
A small islet 0.25 mile offshore of Rosario Beach. Boat access only. CAUTION: there are rocks to the north and west of island.

SHARPE COUNTY PARK (undev)
Located on Rosario Road. The park is undeveloped with no facilities. Parking near the road. The trail to beach is 0.5 mile through trees and moss covered bluffs. CAUTION: the trail is difficult to find and follow, especially near the water and is not recommended for children or others who have difficulty traveling. The trail goes by a large freshwater wetland with many birds and animals. Excellent scenic view from the top of bluffs.

SKAGIT ISLAND STATE PARK (undev)
Boat access only. No facilities, but the island has a very nice camping area on its northeast corner where there is a small beach.

SWINOMISH CHANNEL BOAT LAUNCH
Located under the Highway 20 Bridge over the Swinomish Channel on the east side of the channel. No beach facilities, ample parking.

THOMPSON ROAD ACCESS SITE (undev)
Take Thompson Road south off Highway 20 to a dead end. An undeveloped and unmarked area, but there is a trail to the beach. Limited parking.

SAN JUAN COUNTY

San Juan County, Washington's most marine county, consists of 468 islands (at low tide) which add up to 373 miles of marine shoreline. The San Juans are located at the southern end of the Strait of Georgia, midway between the Washington mainland and Vancouver Island, British Columbia. The four largest and most populated islands—Lopez, Shaw, Orcas, and San Juan—are served by the Washington State Ferry system. The other islands are accessible by private boat. Boat rentals, charters and tours also are available.

Even the most crowded islands of the San Juans are sparsely populated during the winter months. Most people make the San Juans their summer home. The only towns of mentionable size are Friday Harbor on San Juan Island and East Sound on Orcas Island. On summer weekends the towns become quite crowded with tourists who are their main source of business. Some of the more interesting attractions in Friday Harbor are the Whale Museum and the University of Washington Marine Lab. Both are open to visitors all summer and at scheduled times during the winter. Friday Harbor and Roche Harbor are "ports-of-entry" into the United States from Canada and usually filled with boats.

The shore of the islands is generally made up of rocky headlands, pine covered bluffs, crescent beaches, and sheltered coves. Tidepools abundant with intertidal sealife can be found on rocky beaches at low tide, especially at American Camp on

San Juan Island. Many of the small islands and offshore rocks serve as rookeries and support substantial bird and marine mammal populations. The San Juans have the largest concentration of bald eagles outside of Alaska. The area serves as a major wintering area for eagles and also supports a large resident population of both bald and golden eagles.

The rain shadow effect of the Olympics on the San Juans forms a microclimate in which the annual precipitation on some portions of the islands is fifteen inches or less. Cactus can be found growing in these locations within a few feet of marine water. The San Juans enjoy more sunshine and warmer weather than the rest of western Washington; consequently, the islands are popular for recreation.

Many people find that biking is the ideal way to explore the San Juans. Bringing only your bike on the ferry saves you money and the island residents will certainly appreciate a car left behind. Parking at the ferry terminal has been very crowded during the past few years so try to time your trips to avoid peak hours.

Camping is available, but limited, on Orcas, San Juan and Lopez Islands and lodging is also available on these islands. Two large National Historic Parks, English Camp and American Camp, on San Juan Island are informative and interesting to visit. Neither of these large parks allow camping.

If you are lucky enough to be cruising around the San Juan Islands by boat you will have unlimited areas to explore and a good opportunity to view marine life. Many of the islands are part of the San Juan National Wildlife Refuge and are closed to entry in order to protect nesting birds. Boaters are allowed no closer than 100 yards to these islands. Several of the small uninhabited islands are State Parks and offer good camping and moorage. Please be respectful and stay off of private property in the San Juans. There are several islands that may appear isolated but if you go ashore you may have to face an angry landowner.

SAN JUAN COUNTY
ORCAS ISLAND EAST

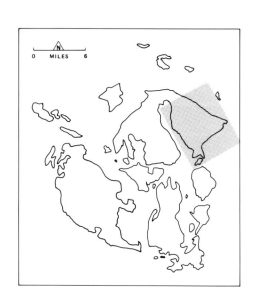

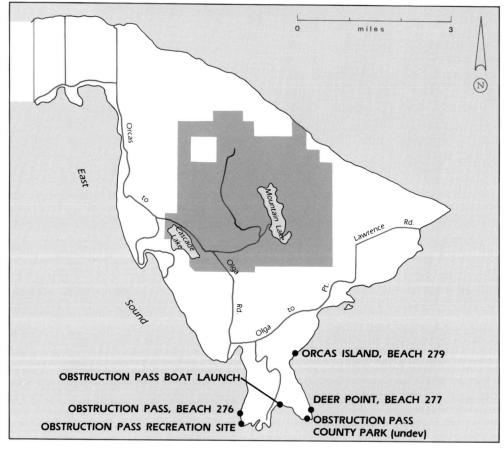

SAN JUAN COUNTY
ORCAS ISLAND EAST

PUBLIC SHORE

	ACRES	CAMP UNITS	PICNIC UNITS	RESTROOMS	FIREPITS	SWIMMING BEACH	BOAT LAUNCH (lanes)	BOAT MOORAGE (slips/buoys)	PUBLIC PIER	DRINKING WATER	VIEWPOINT	SHORELINE LENGTH (feet)	ROCK BEACH	SAND BEACH	GRAVEL BEACH	MUD BEACH	SAND DUNES	TIDE POOLS	WETLANDS	BLUFFS
DEER POINT, BEACH 277	NA											2,700			•					•
OBSTRUCTION PASS BOAT LAUNCH	NA						1					20			•					
OBSTRUCTION PASS COUNTY PARK (undev)	1.0											60	•		•					
OBSTRUCTION PASS RECREATION SITE	10.0	9	2	•	•			2				400			•					•
OBSTRUCTION PASS, BEACH 276	NA											6,941			•					•
ORCAS ISLAND, BEACH 279	NA											1,446			•					•
SAN JUAN ISLANDS NAT'L WILDLIFE REFUGE	460.0	15	15	•	•			3		•	•	40,000	•	•	•					

DEER POINT, BEACH 277
Located on the southeastern tip of Orcas Island. Access is by boat only. This narrow cobble beach has rocky headlands. Only the tidelands are public.

OBSTRUCTION PASS BOAT LAUNCH
Located at the south end of Baylis Road on the southeast tip of Orcas Island. No facilities other than the Boat ramp.

OBSTRUCTION PASS COUNTY PARK (undev)
Located at Deer Point on the southeast tip of Orcas Island. No recreation facilities and access is by boat only.

OBSTRUCTION PASS RECREATION SITE
Located near the south end of the Obstruction Pass Road on the southeast end of Orcas island. Half-mile trail through the woods to a walk-in camp and the beach. A nice crescent beach with pea gravel. Boating access also. Bicycle rack at the parking lot.

OBSTRUCTION PASS, BEACH 276
Located on the eastern portion of Orcas Island, at the southern tip near Obstruction Pass. The site has two distinctly different types of beach: the north end consists of gravel beaches; the south end has rocky headlands with gravel pocket beaches. The Obstruction Pass Recreation Site encompasses most of the uplands and provides upland access.

ORCAS ISLAND, BEACH 279
Located near the southeastern point of Orcas Island. Access is by boat only. The cobble, gravel pocket beaches have rocky headlands. Only the tidelands are public.

SAN JUAN ISLANDS NAT'L WILDLIFE REFUGE
The San Juan Islands National Wildlife Refuge includes 84 rocks, reefs, grassy islands, and forested islands scattered throughout the San Juan Islands. Matia and Turn Islands are the only areas that are open to the public. The remaining islands are set apart and closed to the public to protect colonies of nesting seabirds, including tufted puffins, pigeon guillemots, rhinoceros auklets, cormorants and one of the largest colonies of glaucous-winged gulls on the west coast. They also protect numerous other wildlife. Please view wildlife of closed islands from a distance of at least 200 yards. Please refer to the Fish and Wildlife Service's brochure for the exact location of refuge sites.

SAN JUAN NATIONAL WILDLIFE REFUGE

Eighty-four of the islands scattered throughout the San Juans of northern Puget Sound are designated part of the San Juan Islands National Wildlife Refuge and Wilderness Area. The islands were set aside to protect colonies of nesting seabirds such as tufted puffins, pigeon guillemots, auklets, pelagic cormorants and one of the largest nesting colonies of glaucous-winged gulls on the west coast.

Most of these birds are attracted by the high secluded sea cliffs found on many of the refuge islands. Gulls lay their eggs in the grass at the top of the cliffs, while cormorants prefer a nest of twigs on the cliff ledges. Auklets, puffins, and guillemots nest in burrows or crevices high in the cliff walls.

Other animals also make their home on the islands. Bald eagles find the special protection and seclusion they need for nesting and wintering. Harbor seals frequently rest on the rocks and beaches, while porpoises and whales play and feed along the shore.

To protect the wildlife area, the refuge is managed to keep development and human intrusion to a minimum. The public is asked to view the islands only from a distance. Mooring and picnicking facilities are kept on only two of the larger islands, Matia and Turn, to avoid disturbance to other more sensitive areas. All the remaining islands are closed to the public.

Because strict management of the islands has helped maintain their natural character, all the refuge islands except Smith, Minor, Turn and five acres on Matia Island have been designated as wilderness area. The Wilderness Act of 1964 describes wilderness as "an area where the earth and its community of life are untrammeled by man, where man himself is a visitor who does not remain."

To observe wildlife on the islands, The U.S. Fish and Wildlife Service recommends staying at least 100 yards away from shore. It's a good idea to bring binoculars and a spotting scope to observe with.

SAN JUAN COUNTY
ORCAS ISLAND EAST

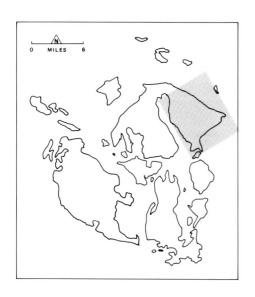
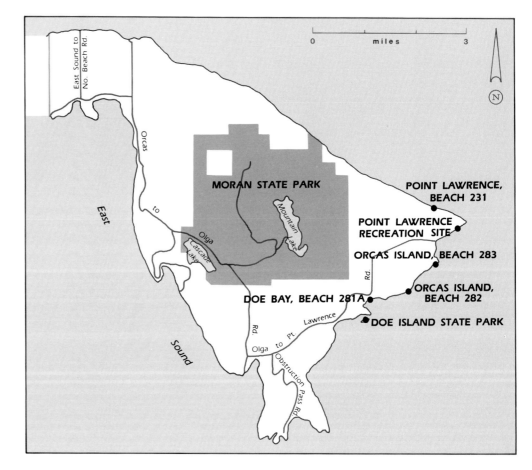

SAN JUAN COUNTY
ORCAS ISLAND EAST

	ACRES	CAMP UNITS	PICNIC UNITS	RESTROOMS	FIREPITS	SWIMMING BEACH	BOAT LAUNCH (lanes)	BOAT MOORAGE (slips/buoys)	PUBLIC PIER	DRINKING WATER	VIEWPOINT	SHORELINE LENGTH (feet)	ROCK BEACH	SAND BEACH	GRAVEL BEACH	MUD BEACH	SAND DUNES	TIDE POOLS	WETLANDS	BLUFFS
DOE BAY, BEACH 281A	NA											1,509	•		•					
DOE ISLAND STATE PARK	6.1	5		•	•			5	•			2,050	•							•
MORAN STATE PARK	4,604.3	148	101	•	•	•	1			•	•	1,800	•							•
ORCAS ISLAND, BEACH 282	NA											1,694	•		•					
ORCAS ISLAND, BEACH 283	NA											3,907	•		•					
POINT LAWRENCE RECREATION SITE (undev)	12.0											536	•	•	•					•
POINT LAWRENCE, BEACH 231	NA											26,386	•	•	•					

DOE BAY, BEACH 281A
Located on Doe Bay on the eastern side of Orcas Island. Access is by boat only. The beach is rocky with patches of gravel and occasional rocks and boulders. Only the tidelands are public.

DOE ISLAND STATE PARK
Located on the southeast side of Orcas Island. Pier and float with room for about 5 boats. No moorage buoys. 2 small crescent beaches. Rocky island with Douglas-fir Forest.

MORAN STATE PARK
Located on Orcas Island. Boat launch is to a freshwater lake. Saltwater shoreline is accessible by boat only. Beach area is mostly unusable. Drive to the top of Mt. Constitution for a commanding view of the San Juans and Mt. Baker, plus other cascade peaks. This large park is primarily designed to be accessible and usable from the roads of Orcas Island and not usable from the water.

ORCAS ISLAND, BEACH 282
Located on the eastern edge of Orcas Island, bordering on Rosario Strait. Access is by boat only. This rocky area has patches of gravel and pea gravel. A few pocket beaches, rocks and boulders are present. Only the tidelands are public.

ORCAS ISLAND, BEACH 283
Located on the eastern edge of Orcas Island, bordering on Rosario Strait and just south of Sea Acre. Access is by boat only. This rocky beach has patches of gravel and pea gravel plus a few pocket beaches with rocks and boulders. Only the tidelands are public.

POINT LAWRENCE RECREATION SITE (undev)
Located at the furthest eastern point of Orcas Island. An undeveloped upland tract which is connected to Beach 231.

POINT LAWRENCE, BEACH 231
Located on the northeastern edge of Orcas Island, bordering on the Strait of Georgia. Access is by boat only. The beach north of Point Lawrence is steep rubble, and south of the point there are pocket beaches. On this southern part, the upper beach is sand or gravel and the outer beach is rocky. Only the tidelands are public, except for the Point Lawrence Recreation Site.

SAN JUAN COUNTY
ORCAS ISLAND EAST

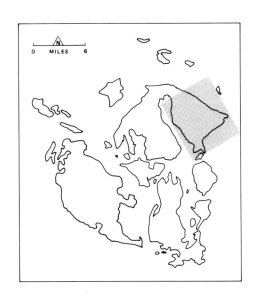
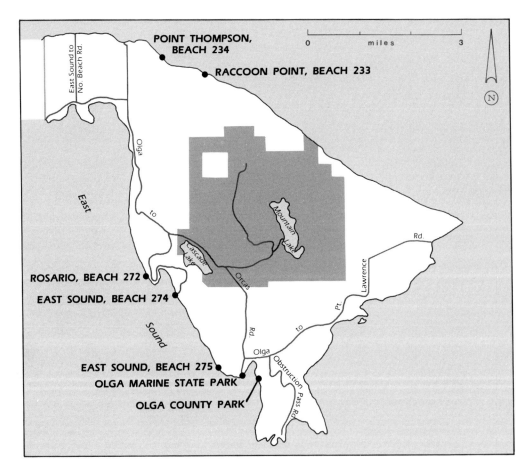

SAN JUAN COUNTY
ORCAS ISLAND EAST

PUBLIC SHORE

	ACRES	CAMP UNITS	PICNIC UNITS	RESTROOMS	FIREPITS	SWIMMING BEACH	BOAT LAUNCH (lanes)	BOAT MOORAGE (slips/buoys)	PUBLIC PIER	DRINKING WATER	VIEWPOINT	SHORELINE LENGTH (feet)	ROCK BEACH	SAND BEACH	GRAVEL BEACH	MUD BEACH	SAND DUNES	TIDE POOLS	WETLANDS	BLUFFS
EAST SOUND, BEACH 274	NA											6,142	•		•					
EAST SOUND, BEACH 275	NA											3,786	•		•					
OLGA COUNTY PARK	1.0							8				30								
OLGA MARINE STATE PARK	1.0							11				60								
POINT THOMPSON, BEACH 234	NA											1,213	•	•	•					
RACCOON POINT, BEACH 233	NA											7,077	•	•	•					
ROSARIO, BEACH 272	NA											4,999	•	•	•					

EAST SOUND, BEACH 274
Located on the eastern side of East Sound near Entrance Mountain on Orcas Island. Access is by boat only. The beach is composed of rocky areas with narrow patches of boulders and pea gravel. Only the tideland areas are public.

EAST SOUND, BEACH 275
Located on the eastern side of East Sound, just west of Olga. Access is by boat only. Tidelands are rocky with narrow boulder and pea gravel beaches. Only the tidelands are public.

OLGA COUNTY PARK
A small area located southeast and across Buck Bay from the town of Olga on Orcas island. No recreation facilities.

OLGA MARINE STATE PARK
Located on the waterfront of the town of Olga on Orcas Island. A dock with no upland facilities. It provides boat moorage. Visitors can walk into town for supplies.

POINT THOMPSON, BEACH 234
Located on the northern edge of Orcas Island, near the mid-section of the island. Access is by boat only. This area, overlooking the Strait of Georgia, is composed of an exposed rocky ledge with patches of sand and gravel. Only the tidelands are public.

RACCOON POINT, BEACH 233
Located on the northern edge of Orcas Island, opposite Matia Island across the Strait of Georgia. Access is by boat only. The exposed rocky ledges have patches of sand and gravel. Only the tidelands are public.

ROSARIO, BEACH 272
Located along East Sound on Orcas Island, just north of Rosario Point. Access is by boat only. The area is rocky with small sand or gravel pocket beaches. Only the tidelands are public.

SAN JUAN COUNTY
ORCAS ISLAND WEST

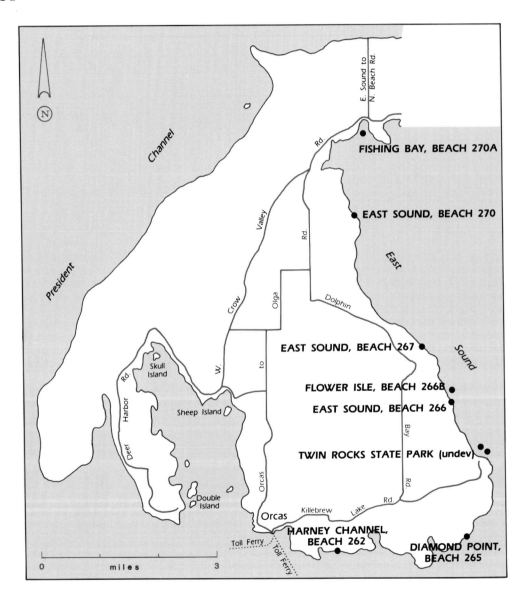

SAN JUAN COUNTY
ORCAS ISLAND WEST

PUBLIC SHORE

	ACRES	CAMP UNITS	PICNIC UNITS	RESTROOMS	FIREPITS	SWIMMING BEACH	BOAT LAUNCH (lanes)	BOAT MOORAGE (slips/buoys)	PUBLIC PIER	DRINKING WATER	VIEWPOINT	SHORELINE LENGTH (feet)	ROCK BEACH	SAND BEACH	GRAVEL BEACH	MUD BEACH	SAND DUNES	TIDE POOLS	WETLANDS	BLUFFS
DIAMOND POINT, BEACH 265	NA											6,685	•		•					
EAST SOUND, BEACH 266	NA											6,155	•		•					
EAST SOUND, BEACH 267	NA											5,024	•		•					
EAST SOUND, BEACH 270	NA											4,790	•	•	•					
FISHING BAY, BEACH 270A	NA											1,240		•	•					
FLOWER ISLE, BEACH 266B	5.0											100	•		•					
HARNEY CHANNEL, BEACH 262	NA											2,203	•		•					
TWIN ROCKS STATE PARK (undev)	NA											300	•		•					

DIAMOND POINT, BEACH 265
Located along East Sound on Orcas Island. Access is by boat only. Diamond Point's rocky headlands have large gravel pockets. Only the tidelands are public.

EAST SOUND, BEACH 266
Located on East Sound in the mid-section of Orcas Island. Access is by boat only. The beach is rocky with narrow boulder and pea gravel sections. Only the tidelands are public.

EAST SOUND, BEACH 267
Located near Dolphin Bay on East Sound in the mid-section of Orcas Island. Access is by boat only. Composition of the beach is rocky with narrow boulder and pea gravel sections. Only the tidelands are public.

EAST SOUND, BEACH 270
One of the many state owned beaches located on the East Sound of Orcas Island. Access is by boat only. The small sand and gravel pocket beaches can be somewhat rocky. Only the tidelands are public.

FISHING BAY, BEACH 270A
A small island 0.25 mile offshore and in the north end of East Sound on Orcas island. Only the tidelands are public.

FLOWER ISLE, BEACH 266B
A small island belonging to the U.S. Bureau of Land Management which is located along the western shore of East Sound on Orcas Island. The beach is rocky.

HARNEY CHANNEL, BEACH 262
Located on the southern edge of Orcas Island, across the water form Pt. Hudson on Shaw Island. Access is by boat only. The beach is rocky with patches of gravel. Only the tidelands are public.

TWIN ROCKS STATE PARK (undev)
Located in the lower west side of East Sound on Orcas Island. No developed recreation facilities. Bird habitat area. Please do not disturb nesting birds.

SAN JUAN COUNTY
ORCAS ISLAND WEST

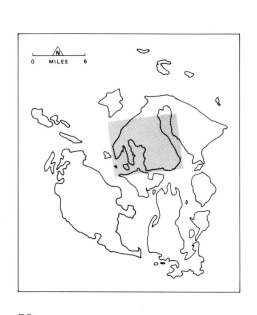
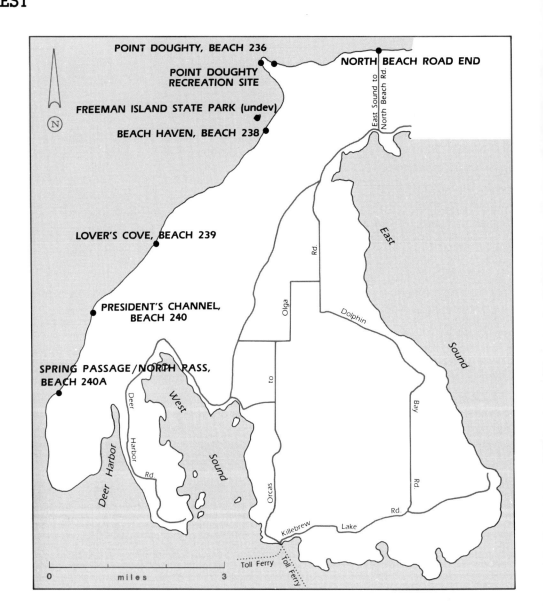

SAN JUAN COUNTY
ORCAS ISLAND WEST

PUBLIC SHORE

	ACRES	CAMP UNITS	PICNIC UNITS	RESTROOMS	FIREPITS	SWIMMING BEACH	BOAT LAUNCH (lanes)	BOAT MOORAGE (slips/buoys)	PUBLIC PIER	DRINKING WATER	VIEWPOINT	SHORELINE LENGTH (feet)	ROCK BEACH	SAND BEACH	GRAVEL BEACH	MUD BEACH	SAND DUNES	TIDE POOLS	WETLANDS	BLUFFS
BEACH HAVEN, BEACH 238	NA											1,201	•		•					
FREEMAN ISLAND STATE PARK (undev)	0.2											100	•		•					
LOVER'S COVE, BEACH 239	NA											2,772		•	•					
NORTH BEACH ROAD END	NA											40			•					
POINT DOUGHTY RECREATION SITE	2.6	2	1	•	•							800	•							
POINT DOUGHTY, BEACH 236	NA											8,256	•							•
PRESIDENT'S CHANNEL, BEACH 240	NA											6,448	•	•	•					
SPRING PASSAGE/NORTH PASS, BEACH 240A	NA											15,813	•		•					•

BEACH HAVEN, BEACH 238
Located on the Northwest shore of Orcas Island. Access is by boat only. The uplands are private. The northern half of this beach is a rocky shelf, the southern half is pea gravel.

FREEMAN ISLAND STATE PARK (undev)
Located 0.25 mile off the northwest shore of Orcas Island 3 miles directly west of Eastsound. No recreation facilities available. Also see Beach 238A.

LOVER'S COVE, BEACH 239
Located on the western edge of Orcas Island, across President Channel from Waldron Island. This stretch of beach has sand and gravel areas that are separated by a rock ledge. Only the tidelands are public.

NORTH BEACH ROAD END
Located at the end of the road to North Beach on the north shore of Orcas Island directly north of Eastsound. Low bank with exposed gravel beach.

POINT DOUGHTY RECREATION SITE
Boating access only. Located at Point Doughty on the northwest shore of Orcas Island. Beach 236 is included as part of this site.

POINT DOUGHTY, BEACH 236
Located on the northern edge of Orcas Island, on a point in the Strait of Georgia. Access is by boat only. The point has rocky headlands and ledge with some cobble beach. This beach is connected to the Point Doughty Recreation Site.

PRESIDENT'S CHANNEL, BEACH 240
Located on the western edge of Orcas Island, across President's Channel from Waldron Island. Access is by boat only. This site is primarily a rock ledge with two or three small pockets of sand or gravel.

SPRING PASSAGE/NORTH PASS, BEACH 240A
Located at the southwestern tip of Orcas Island. Access is by boat only. The area consists of rocky headlands and reefs with sand and gravel pocket beaches. Only the tidelands are public.

SAN JUAN COUNTY
ORCAS ISLAND WEST

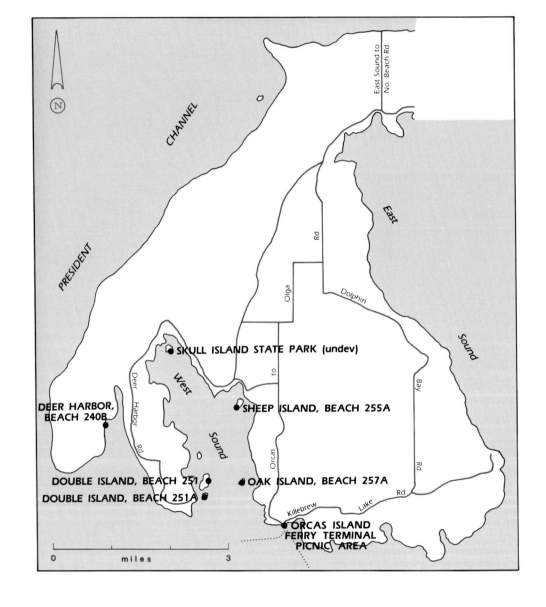
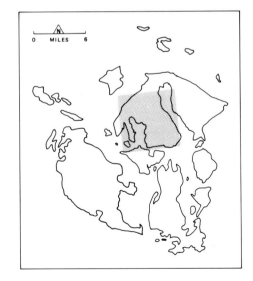

SAN JUAN COUNTY
ORCAS ISLAND WEST

	ACRES	CAMP UNITS	PICNIC UNITS	RESTROOMS	FIREPITS	SWIMMING BEACH	BOAT LAUNCH (lanes)	BOAT MOORAGE (slips/buoys)	PUBLIC PIER	DRINKING WATER	VIEWPOINT	SHORELINE LENGTH (feet)	ROCK BEACH	SAND BEACH	GRAVEL BEACH	MUD BEACH	SAND DUNES	TIDE POOLS	WETLANDS	BLUFFS
DEER HARBOR, BEACH 240B	NA											2,490		•						
DOUBLE ISLAND, BEACH 251	NA											3,960	•	•	•					
DOUBLE ISLAND, BEACH 251A	NA											1,900	•	•	•					
OAK ISLAND, BEACH 257A	NA											300	•	•	•					
ORCAS ISLAND FERRY TERMINAL PICNIC AREA	NA		6	•							•	NA								
SHEEP ISLAND, BEACH 255A	NA											1,081	•	•	•					
SKULL ISLAND STATE PARK (undev)	1.0											NA	•		•			•		•

DEER HARBOR, BEACH 240B
Located in the southwest section of Orcas Island, this beach borders the western edge of Deer Harbor. Access is by boat only. Beach is a narrow rubble area. Only the tidelands are public.

DOUBLE ISLAND, BEACH 251
Located in West Sound just off of Orcas Island. Access is by boat only. The beach surrounding the island is narrow and rocky with sand and gravel patches. Only the tidelands are public.

DOUBLE ISLAND, BEACH 251A
Located in the southern section of Double Island, across from Orcas Island. Access is by boat only. The beach area tends to be narrow, containing rocks plus sand and gravel patches. Only the tidelands are public.

OAK ISLAND, BEACH 257A
A tiny island just off Orcas Island in West Sound. Access is by boat only and can be dangerous. The narrow and rocky beaches have sand and gravel patches or pockets. Only the tidelands are public.

ORCAS ISLAND FERRY TERMINAL PICNIC AREA
Located on Orcas Island immediately uphill from the ferry terminal. View of the ferry terminal. The small day use area is used primarily by people waiting for ferries.

SHEEP ISLAND, BEACH 255A
Sheep Island is located in West Sound off Orcas Island. Access is by boat only. The beach is narrow and rocky with sand and gravel patches or pockets. Only the tidelands are public.

SKULL ISLAND STATE PARK (undev)
Located in West Sound on Orcas Island. This rocky island is accessible by boat only. The park is filled with oak trees. No facilities.

SAN JUAN COUNTY
SUCIA/CLARK ISL.

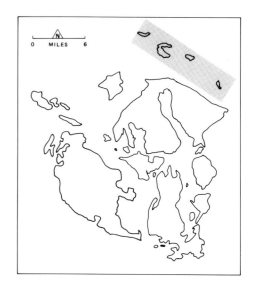

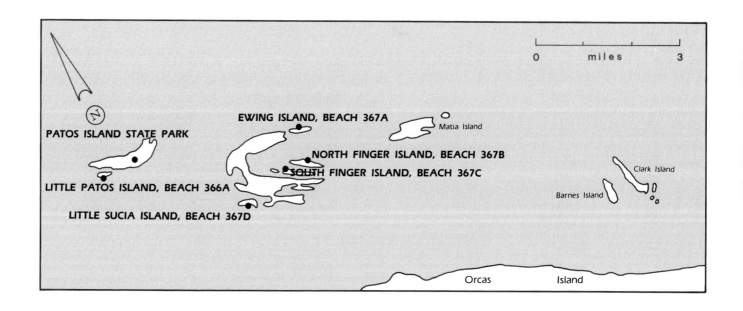

SAN JUAN COUNTY
SUCIA/CLARK ISL.

	ACRES	CAMP UNITS	PICNIC UNITS	RESTROOMS	FIREPITS	SWIMMING BEACH	BOAT LAUNCH (lanes)	BOAT MOORAGE (slips/buoys)	PUBLIC PIER	DRINKING WATER	VIEWPOINT	SHORELINE LENGTH (feet)	ROCK BEACH	SAND BEACH	GRAVEL BEACH	MUD BEACH	SAND DUNES	TIDE POOLS	WETLANDS	BLUFFS
EWING ISLAND, BEACH 367A	NA											5,412	•	•						•
LITTLE PATOS ISLAND, BEACH 366A	NA											3,234	•	•						•
LITTLE SUCIA ISLAND, BEACH 367D	NA											4,930	•	•						•
NORTH FINGER ISLAND, BEACH 367B	NA											8,712	•	•						•
PATOS ISLAND STATE PARK	207.0	4	•	•				2				20,000			•			•		
SOUTH FINGER ISLAND, BEACH 367C	NA											8,375	•	•						•

EWING ISLAND, BEACH 367A
A small island off of the northeastern tip of Sucia Island, which is directly north of the center of Orcas Island. Access is by boat only. The beach is composed of a rocky ledge and headlands with numerous protected sand and gravel pocket beaches.

LITTLE PATOS ISLAND, BEACH 366A
A small island located next to Patos Island, both of which are north of Orcas Island. Access is by boat only. The beach has rocky ledges and headlands with numerous protected sand and gravel pocket beaches. Only the tidelands are public.

LITTLE SUCIA ISLAND, BEACH 367D
A small island just off of Sucia Island, both of which are located north of Orcas Island. Access is by boat only. The beach is made up of rocky ledges and headlands with numerous protected sand and gravel pockets. Only the tidelands are public.

NORTH FINGER ISLAND, BEACH 367B
One of the small islands in Echo Bay just off of Sucia Island. Access is by boat only. The site is composed of rocky ledges and headlands with numerous protected sand and gravel pocket beaches. Only the tidelands are public.

PATOS ISLAND STATE PARK
The island has protected beaches dispersed between rocky bluffs, an old light house and 2 moorage buoys. Unique rock formations. No fresh water. The park includes all of Patos island and is accessible by boat only.

SOUTH FINGER ISLAND, BEACH 367C
Located in Echo Bay adjacent to Sucia Island. Access is by boat only. This tiny island has rocky ledges and headlands with numerous protected sand and gravel pocket beaches. Only the tidelands to the public.

SAN JUAN COUNTY
SUCIA/CLARK ISL.

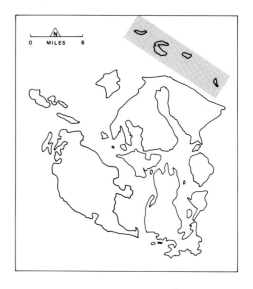

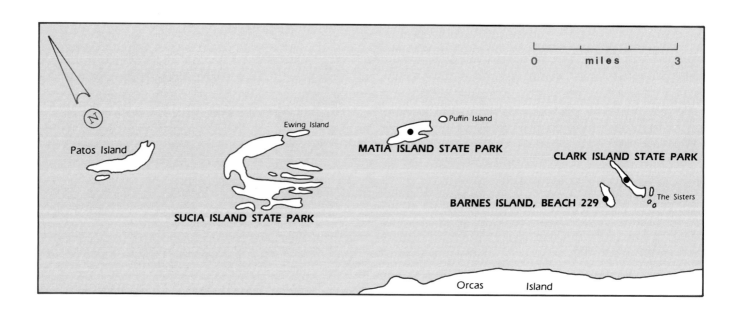

SAN JUAN COUNTY
SUCIA/CLARK ISL.

	ACRES	CAMP UNITS	PICNIC UNITS	RESTROOMS	FIREPITS	SWIMMING BEACH	BOAT LAUNCH (lanes)	BOAT MOORAGE (slips/buoys)	PUBLIC PIER	DRINKING WATER	VIEWPOINT	SHORELINE LENGTH (feet)	ROCK BEACH	SAND BEACH	GRAVEL BEACH	MUD BEACH	SAND DUNES	TIDE POOLS	WETLANDS	BLUFFS
BARNES ISLAND, BEACH 229	NA											6,457	●	●	●					
CLARK ISLAND STATE PARK	55.1	8	2	●	●			8				6,000	●	●	●				●	●
MATIA ISLAND STATE PARK	145.0	6		●	●			6				20,000	●		●				●	●
SUCIA ISLAND STATE PARK	562.1	51		●	●			74		●	●	45,000	●	●	●			●	●	●

BARNES ISLAND, BEACH 229
Located 1.0 mile off the northeast shore of Orcas Island. Access is by boat only. The entire tidal shore of the island is public. The beach is a rocky ledge with sections of sand and gravel.

CLARK ISLAND STATE PARK
Located northeast of Orcas Island. The entire island is state owned. Unprotected moorage and two extensive gravel/sand beaches.

MATIA ISLAND STATE PARK
Boat access only. Campsites are in protected coves on the west side of the island. Rocky bluffs surround the island. There are many protected coves and beautiful geologic formations. Lots of marine life and birds, including bald eagles.

SUCIA ISLAND STATE PARK
Located 2.5 miles north of Orcas Island, Sucia Island offers camping and picnic areas spread throughout the island. 6 miles of hiking trails allow for exploration of this uniquely shaped island. There are 2 docks in Fossil Bay and 16 buoys in Echo Bay. The island has many habitat types, from forests of juniper and Pacific yew trees to rock, sand and gravel shores. Water is available. Toilet tank disposal.

SAN JUAN COUNTY
STUART/WALDRON ISL.

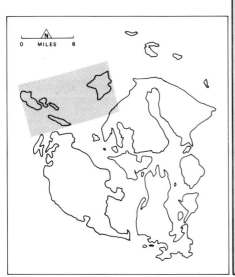

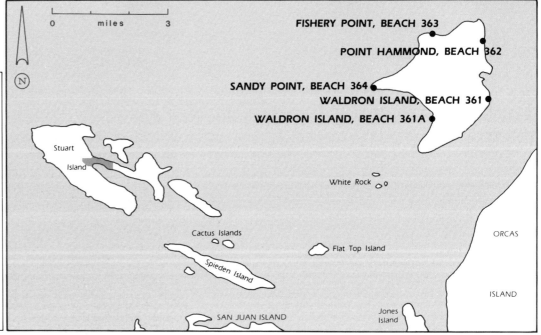

SAN JUAN COUNTY
STUART/WALDRON ISL.

	ACRES	CAMP UNITS	PICNIC UNITS	RESTROOMS	FIREPITS	SWIMMING BEACH	BOAT LAUNCH (lanes)	BOAT MOORAGE (slips/buoys)	PUBLIC PIER	DRINKING WATER	VIEWPOINT	SHORELINE LENGTH (feet)	ROCK BEACH	SAND BEACH	GRAVEL BEACH	MUD BEACH	SAND DUNES	TIDE POOLS	WETLANDS	BLUFFS
FISHERY POINT, BEACH 363	NA											4,422	•		•					
POINT HAMMOND, BEACH 362	NA											3,218	•		•					
SANDY POINT, BEACH 364	NA											2,640	•		•					
WALDRON ISLAND, BEACH 361	NA											4,950			•					
WALDRON ISLAND, BEACH 361A	NA											1,467	•		•					

FISHERY POINT, BEACH 363
Located on the northern edge of Waldron Island, an island to the northwest of Orcas Island. Access is by boat only. The beach is composed of gravel and sand with scattered boulders. Only the tidelands are public.

POINT HAMMOND, BEACH 362
Located on the eastern edge of Waldron Island, just across from Orcas Island. Access is by boat only. Contains sand and gravel beaches with scattered boulders. Only the tidelands are public.

SANDY POINT, BEACH 364
Located on the western edge of Waldron Island, along the northern and southern edges of Sandy Point. Access is by boat only. Sand and gravel beaches are scattered with boulders. Only the tidelands are public.

WALDRON ISLAND, BEACH 361
Located on the eastern edge of Waldron Island which is northwest of Orcas Island. Access is by boat only. The narrow rubble beach has some gravelly areas. Only the tidelands are public.

WALDRON ISLAND, BEACH 361A
Located in Cowlitz Bay on the southwestern side of Waldron Island. There is upland access to the beach. The narrow rubble beach has some gravelly areas. The dock is owned by San Juan County. Only the tidelands are public.

SAN JUAN COUNTY
STUART/WALDRON ISL.

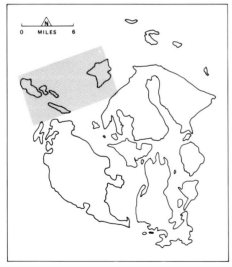

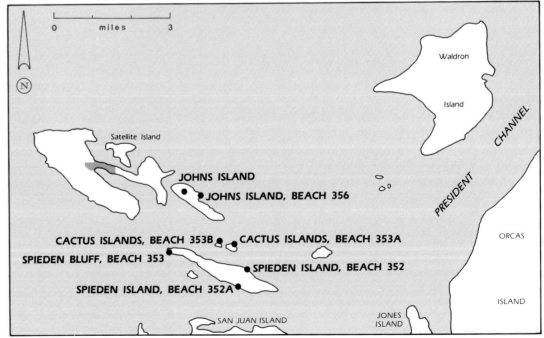

SAN JUAN COUNTY
STUART/WALDRON ISL.

PUBLIC SHORE

	ACRES	CAMP UNITS	PICNIC UNITS	RESTROOMS	FIREPITS	SWIMMING BEACH	BOAT LAUNCH (lanes)	BOAT MOORAGE (slips/buoys)	PUBLIC PIER	DRINKING WATER	VIEWPOINT	SHORELINE LENGTH (feet)	ROCK BEACH	SAND BEACH	GRAVEL BEACH	MUD BEACH	SAND DUNES	TIDE POOLS	WETLANDS	BLUFFS
CACTUS ISLANDS, BEACH 353A	NA											2,523	•	•	•					
CACTUS ISLANDS, BEACH 353B	NA											4,150	•							
JOHNS ISLAND	4.3											1,000			•					
JOHNS ISLAND, BEACH 356	NA											20,368			•				•	
SPIEDEN BLUFF, BEACH 353	NA											1,054	•							
SPIEDEN ISLAND, BEACH 352	NA											1,458	•							
SPIEDEN ISLAND, BEACH 352A	NA											6,460	•							

CACTUS ISLANDS, BEACH 353A
Includes all the tidelands of Cactus Islands. The uplands are private. The beaches are gravel or sand over a rocky shelf.

CACTUS ISLANDS, BEACH 353B
Includes all the tidelands of the smaller Cactus island. The uplands are private. The beach is rocky.

JOHNS ISLAND
A small parcel on the northeast shore of Johns Island 0.5 mile southeast of Johns Pass, east of Stuart Island. Access is by boat only. Undeveloped with no facilities.

JOHNS ISLAND, BEACH 356
Johns Island is one of the islands located north of San Juan Island. Access is by boat only. The beaches which surround the island are pea gravel with rocky headlands. Only the tidelands are public.

SPIEDEN BLUFF, BEACH 353
Located on Spieden Island, just north of San Juan Island. Access is by boat only. The beach is rocky. Only the tidelands are public.

SPIEDEN ISLAND, BEACH 352
This beach is located on the northern edge of Spieden Island, a long and narrow island located north of San Juan Island. Access is by boat only. The beach is rocky. Only the tidelands are public.

SPIEDEN ISLAND, BEACH 352A
Located along the southern edge of Spieden Island, bordering on Spieden Channel across from the northern edge of San Juan Island. Access is by boat only. The beach is rocky. Only the tidelands are public.

SAN JUAN COUNTY
STUART/WALDRON ISL.

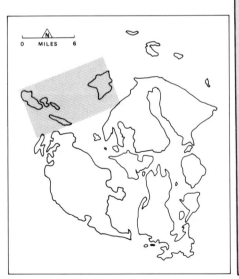

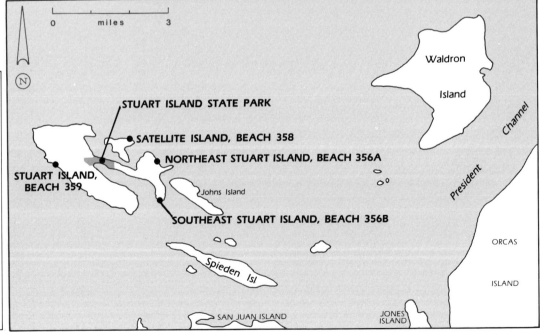

SAN JUAN COUNTY
STUART/WALDRON ISL.

	ACRES	CAMP UNITS	PICNIC UNITS	RESTROOMS	FIREPITS	SWIMMING BEACH	BOAT LAUNCH (lanes)	BOAT MOORAGE (slips/buoys)	PUBLIC PIER	DRINKING WATER	VIEWPOINT	SHORELINE LENGTH (feet)	ROCK BEACH	SAND BEACH	GRAVEL BEACH	MUD BEACH	SAND DUNES	TIDE POOLS	WETLANDS	BLUFFS
NORTHEAST STUART ISLAND, BEACH 356A	NA											8,780	•	•						•
SATELLITE ISLAND, BEACH 358	NA											14,295	•	•						•
SOUTHEAST STUART ISLAND, BEACH 356B	NA											3,792	•							
STUART ISLAND STATE PARK	152.6	19	•	•				48		•	•	5,130			•					•
STUART ISLAND, BEACH 359	NA											36,050	•	•						•

NORTHEAST STUART ISLAND, BEACH 356A
Located near Johns Pass on Stuart Island, which is northwest of San Juan Island. Access is by boat only. These rocky headlands have rocky ledges with sand and gravel pocket beaches. Only the tidelands are public.

SATELLITE ISLAND, BEACH 358
Located around the perimeter of Satellite Island, which is tucked into Stuart Island north of San Juan Island. Access is by boat only. The area consists of rocky headlands and rocky ledge with sand or gravel pocket beaches. Only the tidelands are public.

SOUTHEAST STUART ISLAND, BEACH 356B
Located directly across from Johns Island in the waters north of San Juan Island. Access is by boat only. The beach is rocky. Only the tidelands are public.

STUART ISLAND STATE PARK
Located 2 miles northwest of San Juan Island, Stuart Island is the westernmost of the San Juan Islands group. The state park is near the center of the island and may be approached either from Reid Harbor on the south or Prevost Harbor on the north. Access is by boat only. There are trails throughout the island. Two "floating islands" are available for docking at the head of Reid Harbor.

STUART ISLAND, BEACH 359
Located along much of the northern and western edge of Stuart Island, running in both directions from Turn Point. Access is by boat only. Composition of this extensive area is rocky headlands and rocky ledges with sand or gravel pocket beaches. Only the tidelands are public.

SAN JUAN COUNTY
SAN JUAN ISLAND

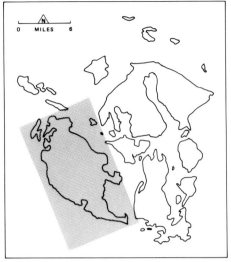
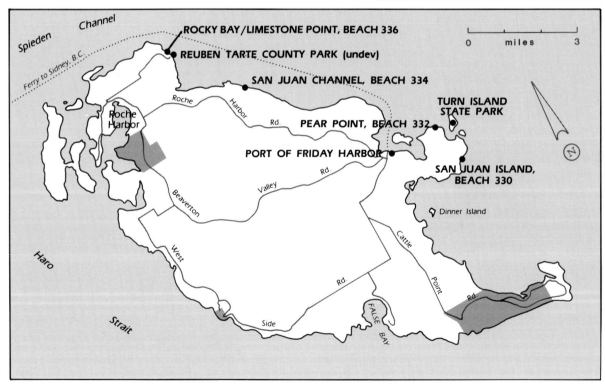

SAN JUAN COUNTY
SAN JUAN ISLAND

	ACRES	CAMP UNITS	PICNIC UNITS	RESTROOMS	FIREPITS	SWIMMING BEACH	BOAT LAUNCH (lanes)	BOAT MOORAGE (slips/buoys)	PUBLIC PIER	DRINKING WATER	VIEWPOINT	SHORELINE LENGTH (feet)	ROCK BEACH	SAND BEACH	GRAVEL BEACH	MUD BEACH	SAND DUNES	TIDE POOLS	WETLANDS	BLUFFS
PEAR POINT, BEACH 332	NA		1									6,097	•	•					•	
PORT OF FRIDAY HARBOR	1.2		•					250	•			800			•					
REUBEN TARTE COUNTY PARK (undev)	6.0										•	600	•		•			•		
ROCKY BAY/LIMESTONE POINT, BEACH 336	NA											11,827			•					
SAN JUAN CHANNEL, BEACH 334	NA											14,516			•				•	
SAN JUAN ISLAND, BEACH 330	NA											8,432	•	•					•	
TURN ISLAND STATE PARK	35.2	10	1	•	•			3				16,000			•			•		

PEAR POINT, BEACH 332
Located on San Juan Island, just east of Friday Harbor. Access is by boat only. The area contains rocky headlands and ledges with sand and gravel pocket beaches. Only the tidelands are public.

PORT OF FRIDAY HARBOR
Located in downtown Friday Harbor north of the ferry dock. Recently developed park with complete boating facilities. Customs office. Overnight moorage, showers and restrooms available. Stores and restaurants within walking distance.

REUBEN TARTE COUNTY PARK (undev)
Located on northern San Juan Island. Take Roche Harbor Road to Rouleau Road; turn on Limestone Point Road and go to San Juan Drive; in 0.5 mile turn left on a dirt road at a "no camping" sign. This is a small rocky point near Limestone Point. The very steep gravel road to the park is not recommended for RV's. No developed facilities for recreational use.

ROCKY BAY/LIMESTONE POINT, BEACH 336
Located on San Juan Island just south of Limestone Point on the northeastern shoreline. Access is by boat only. The area consists of rocky headlands with gravel pocket beaches. Only the tidelands are public.

SAN JUAN CHANNEL, BEACH 334
Located along the northeastern edge of San Juan Island. Access is by boat only. There are rocky headlands with a beach composed of pea gravel pockets. Only the tidelands are public.

SAN JUAN ISLAND, BEACH 330
Located on the eastern edge of San Juan Island near Pear Point. Access is by boat only. Contains rocky headlands or ledges with sand or gravel pocket beaches. Only the tidelands are public.

TURN ISLAND STATE PARK
Boating access only. A small forested island off San Juan Island near Friday Harbor. Moorage buoys are available. There are trails around the island.

SAN JUAN COUNTY
SAN JUAN ISLAND

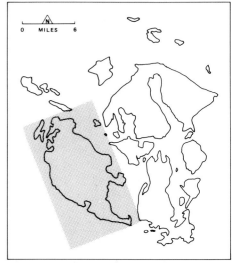

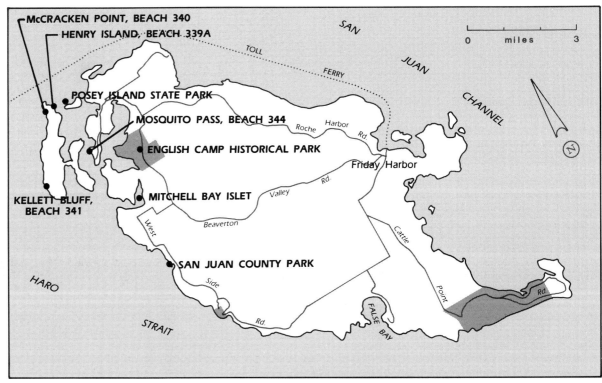

SAN JUAN COUNTY
SAN JUAN ISLAND

	ACRES	CAMP UNITS	PICNIC UNITS	RESTROOMS	FIREPITS	SWIMMING BEACH	BOAT LAUNCH (lanes)	BOAT MOORAGE (slips/buoys)	PUBLIC PIER	DRINKING WATER	VIEWPOINT	SHORELINE LENGTH (feet)	ROCK BEACH	SAND BEACH	GRAVEL BEACH	MUD BEACH	SAND DUNES	TIDE POOLS	WETLANDS	BLUFFS
ENGLISH CAMP HISTORICAL PARK	529.0	10	•							•	•	7,920	•		•			•		
HENRY ISLAND, BEACH 339A	NA											2,408			•					•
KELLETT BLUFF, BEACH 341	NA											11,933	•		•					
McCRACKEN POINT, BEACH 340	NA											2,891			•					
MITCHELL BAY ISLET	2.0											NA			•	•				
MOSQUITO PASS, BEACH 344	NA											488			•					
POSEY ISLAND STATE PARK	1.0	1	2	•								NA	•		•					
SAN JUAN COUNTY PARK	15.0	11		•			1	2				NA	•		•			•		

ENGLISH CAMP HISTORICAL PARK
From Friday Harbor take Beaverton Valley Road to West Valley Road. This is a day use only area with a historical interpretive center that gives slide shows. Hiking trails. Ranger is on the site only in the summer.

HENRY ISLAND, BEACH 339A
The beach on Henry Island is located near the northwestern tip of San Juan Island, facing Roche Harbor. Access is by boat only. The tideland area is made up of gravel pocket beaches with rocky headlands. Only the tidelands are public.

KELLETT BLUFF, BEACH 341
Located on the outer southwestern portion of Henry Island which is off the northwest shore of San Juan Island. Access is by boat only. The bluff is rocky, with a small gravel beach on the north end. Only the tidelands are public.

McCRACKEN POINT, BEACH 340
Located on the northern tip of Henry Island which is just northwest of San Juan Island. Access is by boat only. Site consists of gravel beaches. Only the tidelands are public.

MITCHELL BAY ISLET
A small islet near the shore at the east end of Mitchell Bay on the west side of San Juan Island. Extensive mud flats exposed at low tide. The island is owned by the U.S. Bureau of Land Management, but it is surrounded by privately owned tidelands.

MOSQUITO PASS, BEACH 344
Located on the western side of San Juan Island, near White Point just south of Roche Harbor. Access is by boat only. The beaches are composed of gravel. Only the tidelands are public.

POSEY ISLAND STATE PARK
Located e mile northwest of San Juan Island's Roche Harbor. Accessible by private boat only. Limited facilities. This tiny park has only one campsite.

SAN JUAN COUNTY PARK
Located on the west side of San Juan Island. From Friday Harbor take Beaverton Valley Road west to West Valley Road; turn left on Mitchell Bay Road and then left on West Side Road. Tidepools and rocky shores to explore. Primitive camping facilities. This is the only camping on San Juan Island.

SAN JUAN COUNTY
SAN JUAN ISLAND

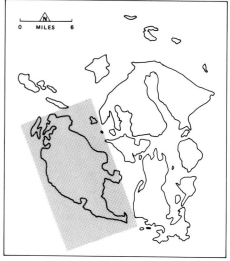

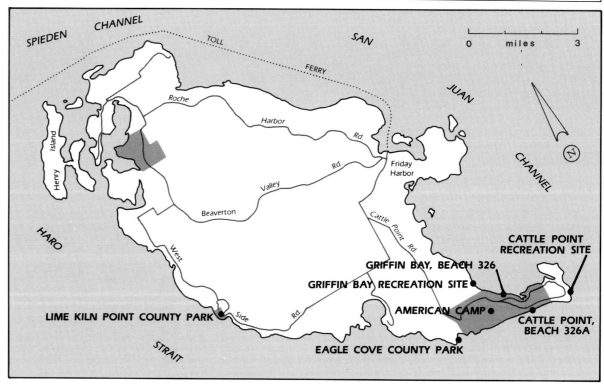

SAN JUAN COUNTY
SAN JUAN ISLAND

	ACRES	CAMP UNITS	PICNIC UNITS	RESTROOMS	FIREPITS	SWIMMING BEACH	BOAT LAUNCH (lanes)	BOAT MOORAGE (slips/buoys)	PUBLIC PIER	DRINKING WATER	VIEWPOINT	SHORELINE LENGTH (feet)	ROCK BEACH	SAND BEACH	GRAVEL BEACH	MUD BEACH	SAND DUNES	TIDE POOLS	WETLANDS	BLUFFS
AMERICAN CAMP	1,223.0		20	•							•	22,440	•		•			•	•	•
CATTLE POINT LIGHTHOUSE RECREATION SITE	8.2		8	•	•				•	•		6,600	•	•				•		•
CATTLE POINT, BEACH 326A	NA											28,242	•	•						
EAGLE COVE COUNTY PARK	0.3	5	3				2					100		•						
GRIFFIN BAY RECREATION SITE	15.0	1	3	•	•		2			•		327		•						
GRIFFIN BAY, BEACH 326	NA											3,498	•	•						
LIME KILN POINT COUNTY PARK	39.1			•							•	2,550	•					•		

AMERICAN CAMP
Located on San Juan Island; take Cattle Point Road to American Camp Road. This historic park includes an interpretive trail. Notice the open desert like area; portions of the San Juan Islands are very dry. Many birds to observe.

CATTLE POINT LIGHTHOUSE RECREATION SITE
From Friday Harbor take Cattle Point Road through American Camp. Walk to the lighthouse. Includes an interpretive area, birds and tidepools. Provides access to Cattle Point, Beach 326A.

CATTLE POINT, BEACH 326A
Encompasses most of the tidelands surrounding the southeastern point of San Juan island. Upland access is through the DNR's Cattle Point Recreation Site and from the national park property of the San Juan National Historic Park. The remaining uplands are private. The area has mostly wide beaches of sand and gravel except for rocky headlands and pocket beaches north of Cattle Point.

EAGLE COVE COUNTY PARK
A small beach area which borders on the San Juan National Historic Park, but access is through a private subdivision with no developed parking.

GRIFFIN BAY RECREATION SITE
Boating access only. Toilets are up the road from the beach. Water is available. Picnic areas by the water.

GRIFFIN BAY, BEACH 326
Located on San Juan Island along Griffin Bay, northwest of the Cattle Point Recreation Site. Accessible by walking across national park property. The beaches are wide sand and gravel areas. Only the tidelands are owned by the state.

LIME KILN POINT COUNTY PARK
From Friday Harbor take San Juan Valley Road; go left on Douglas Road, right on Bailer Hill Road to West Side Road; after West Side Road becomes a gravel road for 0.5 mile, park on left. Trail to the lighthouse and Point is down the gravel road.

SAN JUAN COUNTY
LOPEZ ISLAND

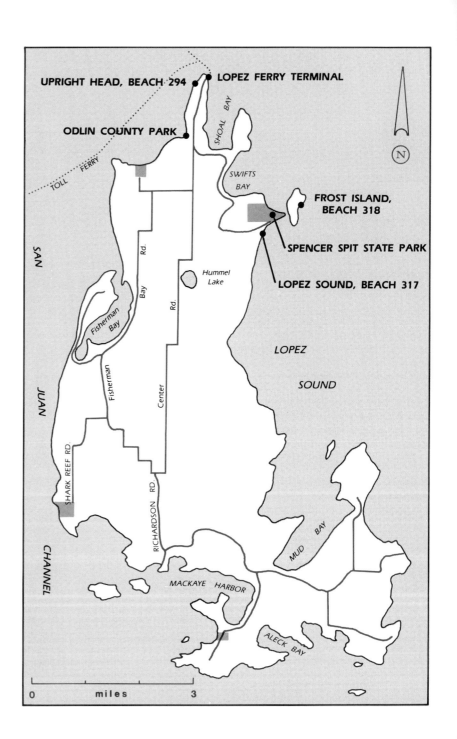

SAN JUAN COUNTY
LOPEZ ISLAND

PUBLIC SHORE

	ACRES	CAMP UNITS	PICNIC UNITS	RESTROOMS	FIREPITS	SWIMMING BEACH	BOAT LAUNCH (lanes)	BOAT MOORAGE (slips/buoys)	PUBLIC PIER	DRINKING WATER	VIEWPOINT	SHORELINE LENGTH (feet)	ROCK BEACH	SAND BEACH	GRAVEL BEACH	MUD BEACH	SAND DUNES	TIDE POOLS	WETLANDS	BLUFFS
FROST ISLAND, BEACH 318	NA											6,138	●							●
LOPEZ FERRY TERMINAL	NA		3	●						●		NA	●					●		
LOPEZ SOUND, BEACH 317	NA											11,773			●					
ODLIN COUNTY PARK	85.0	16	6	●	●		1					7,200		●						
SPENCER SPIT STATE PARK	129.4	46	15	●	●	●		16		●	●	7,840		●	●				●	
UPRIGHT HEAD, BEACH 294	NA											6,794			●					●

FROST ISLAND, BEACH 318
Located off the northeast shore of Lopez Island at Spencer spit. Access is by boat only. The island is steep, and rocky with rocky beaches *and not usable except for a small rock beach at very low tide. Only the tidelands are public.*

LOPEZ FERRY TERMINAL
The area at the ferry terminal provides parking, a sheltered waiting area and a booth for ferry information. There is very limited beach access for those who wish to climb down a steep bank.

LOPEZ SOUND, BEACH 317
Located along the northwestern edge of Lopez Island. Access is available through Spencer Spit State Park. The upper beach is cobble and scattered boulders, whereas the lower beach is cobble and sand. Only the tidelands are public.

ODLIN COUNTY PARK
Located on Lopez Island 1 mile from the ferry dock; go right on Ferry Road when driving away from the ferry. Protected sandy beach.

SPENCER SPIT STATE PARK
Located on Lopez Island 5 miles from the ferry terminal. The salt marsh lagoon in the park is home to innumerable forms of aquatic life and sea and shore birds. This park is newly developed and nicely done. Watch the state ferries and other vessels in the main commercial channel just to the north. Sloping uplands are forested.

UPRIGHT HEAD, BEACH 294
Located on the northern tip of Lopez Island. Access is available from the ferry terminal. The headlands have a gravel pocket beach located near the ferry terminal. Only the tidelands are public.

SAN JUAN COUNTY
LOPEZ ISLAND

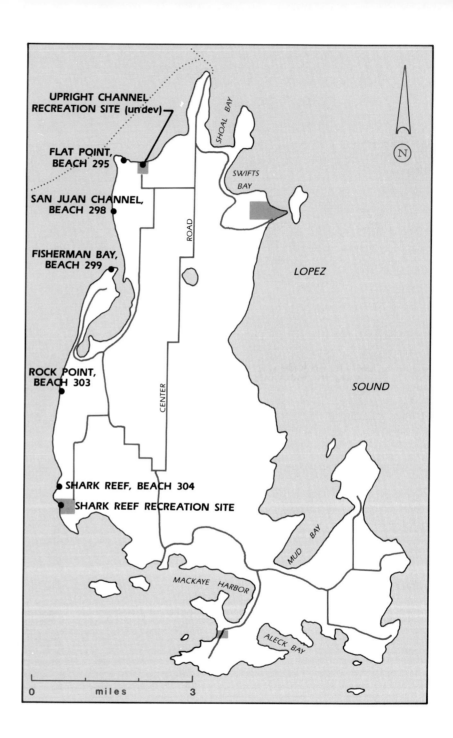

SAN JUAN COUNTY
LOPEZ ISLAND

	ACRES	CAMP UNITS	PICNIC UNITS	RESTROOMS	FIREPITS	SWIMMING BEACH	BOAT LAUNCH (lanes)	BOAT MOORAGE (slips/buoys)	PUBLIC PIER	DRINKING WATER	VIEWPOINT	SHORELINE LENGTH (feet)	ROCK BEACH	SAND BEACH	GRAVEL BEACH	MUD BEACH	SAND DUNES	TIDE POOLS	WETLANDS	BLUFFS
FISHERMAN BAY, BEACH 299	NA											1,660	•		•					
FLAT POINT, BEACH 295	NA											6,081			•					
ROCK POINT, BEACH 303	NA											1,312			•					
SAN JUAN CHANNEL, BEACH 298	NA											5,347	•		•					
SHARK REEF RECREATION SITE	38.1											NA	•		•				•	
SHARK REEF, BEACH 304	NA											13,965	•		•					
UPRIGHT CHANNEL RECREATION SITE (undev)	20.4											700			•					

FISHERMAN BAY, BEACH 299
Located on Lopez Island near the town of Lopez. Access is by boat only. The beach is sand or gravel. Only the tidelands are public.

FLAT POINT, BEACH 295
Located at Flat Point on Lopez Island. Accessible by road through the Upright Channel Recreation Site (undeveloped) or from Odlin County Park. Flat Point has a nice gravel beach. Only the tidelands are public.

ROCK POINT, BEACH 303
Located on the western edge of Lopez Island, just south of Fisherman Bay. Access is by boat only. Beach composition is cobble with pea gravel on upper beach, and scattered boulders. Only the tidelands are public.

SAN JUAN CHANNEL, BEACH 298
Located on the western edge of Lopez Island. Access is by boat only. This cobble beach has sandy areas with scattered boulders. Only the tidelands are public.

SHARK REEF RECREATION SITE
Located at the end of the Shark Reef Road on Lopez Island. Look for a path that leads into the park. Sea lions can be seen on the rocks.

SHARK REEF, BEACH 304
Located along the western edge of Lopez Island, running north from Davis Point. Access to the rocky tidelands is available by the access road to the Shark Reef Recreation Site. Extensive gravel or cobble beaches on both ends of the recreation site. The central part of the beach, from Kings Point to Davis Point, is a rocky ledge with small pocket beaches. Only the tidelands are public.

UPRIGHT CHANNEL RECREATION SITE (undev)
Located on the northeast corner of Lopez Island. Connects with Beach 295. Undeveloped with no facilities.

SAN JUAN COUNTY
LOPEZ ISLAND

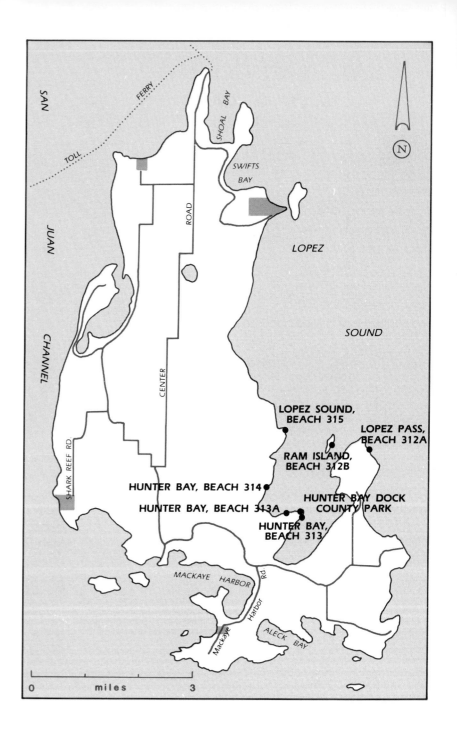

SAN JUAN COUNTY
LOPEZ ISLAND

	ACRES	CAMP UNITS	PICNIC UNITS	RESTROOMS	FIREPITS	SWIMMING BEACH	BOAT LAUNCH (lanes)	BOAT MOORAGE (slips/buoys)	PUBLIC PIER	DRINKING WATER	VIEWPOINT	SHORELINE LENGTH (feet)	ROCK BEACH	SAND BEACH	GRAVEL BEACH	MUD BEACH	SAND DUNES	TIDE POOLS	WETLANDS	BLUFFS
HUNTER BAY DOCK COUNTY PARK	1.0						1					100	•	•						
HUNTER BAY, BEACH 313	NA											2,098			•					•
HUNTER BAY, BEACH 313A	NA											1,300	•		•					
HUNTER BAY, BEACH 314	NA											3,366			•					•
LOPEZ PASS, BEACH 312A	NA											5,913			•					•
LOPEZ SOUND, BEACH 315	NA											9,066			•					•
RAM ISLAND, BEACH 312B	NA											3,890			•					•

HUNTER BAY DOCK COUNTY PARK
Located on Lopez Island on the point between Hunter and Mud bays. From Mud Bay Road turn east on Islandale Road. There is a boat launch in addition to the dock.

HUNTER BAY, BEACH 313
Located on Lopez Island at the point midway between Hunter Bay and Mud Bay. Access is via county owned dock and boat ramp on the west end. The area consists of rocky headlands with a gravelly cobble beach. Only the tidelands are public.

HUNTER BAY, BEACH 313A
Located on Lopez Island on the eastern side of Hunter Bay. Access is by boat only. Beach composition is cobble and coarse sand. Only the tidelands are public.

HUNTER BAY, BEACH 314
Located on Lopez Island on the northern side of Hunter Bay. Access by boat only. Typically rocky headlands with rock reefs and pea gravel pocket beaches. Only the tidelands are public.

LOPEZ PASS, BEACH 312A
Located on Lopez Island across Lopez Pass from Decatur Island. Access is by boat only. This area features rocky headlands and narrow gravelly beaches. Only the tidelands are public.

LOPEZ SOUND, BEACH 315
Located on the eastern side of Lopez Island, north of Jasper Bay. Access is by boat only. The headlands tend to be rocky and the beaches are pea gravel pockets. Only the tidelands are public.

RAM ISLAND, BEACH 312B
A small island off the eastern edge of Lopez, bordering on Lopez Pass. Access is by boat only. The beach is a rocky ledge with two small pockets of gravel. Only the tidelands are public.

ORCAS

To other marine animals, orcas, or killer whales, are fearsome predators who hunt and eat whatever they desire. To coastal tribes like the Makah, orcas were the ultimate challenge to the hunter. Today, many people are awed and fascinated by the intelligence, playfulness, power, and grace of this giant marine mammal.

Washington's waters are one of the few areas of the world where families of orcas successfully live near human populations. Three pods (family groups) of orcas totaling about eighty whales make their home in northern Puget Sound. Several other transient pods visit Puget Sound and the straits, but they rarely stay for long.

Orcas are often seen from ferries, other boats, or from high bluffs around the San Juans and northern Puget Sound. They are most easily identified from a distance by their huge dorsal (back) fin which, in males, can be six feet high and three feet wide at the base. The fins will calmly slice the surface of the water as orcas rise for air, then slip back in, leaving no trace of the five-ton creature below the surface. As one gets closer to an orca, it's easy to spot its black glossy back which contrasts sharply with its flashy white undersides.

Mature males can reach thirty feet and females are slightly smaller. To hear the soft "pouff" of an orca surfacing for air, or to witness the explosion of water as an orca breaches is as frightening as it is exciting.

Unlike most large whales, orcas have teeth which they use for hunting and eating prey (hence the name "killer whale"). Because of their teeth they are considered members of the family Delphinidae, which includes other toothed whales and dolphins. They often hunt in "wolf pack" fashion and it is suspected they usually travel and feed at night. Though they occasionally feed on sea lions, seals, and porpoises, Puget Sound orcas are primarily piscivorous: they eat all types of fish, but like most of us they prefer salmon. It is estimated that they can eat up to 250 pounds of salmon per day.

Orcas live and travel in matriarchal family groups called pods for the duration of their lives, which may span a century. Each pod has its own traveling, feeding and communication patterns. Scientists are currently studying the different dialects of the resident pods in Puget Sound.

The Moclips Cetological Society has a whale research station and public education program. The society runs the Whale Museum, located within walking distance of the ferry dock in Friday Harbor, which has a wealth of information presented in a creative manner on all types of whales. The museum has several whale skeletons, video displays, recordings of whales communicating, and special exhibits for children, among many other features. It also has extensive information on the orca pods that live around the islands. The museum is an attraction that should not be missed by anyone who is interested in marine mammals. The museum requests that whale sightings be reported to 1-800-562-8832.

SAN JUAN COUNTY
LOPEZ ISLAND

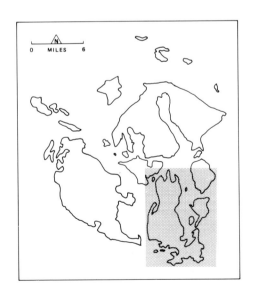
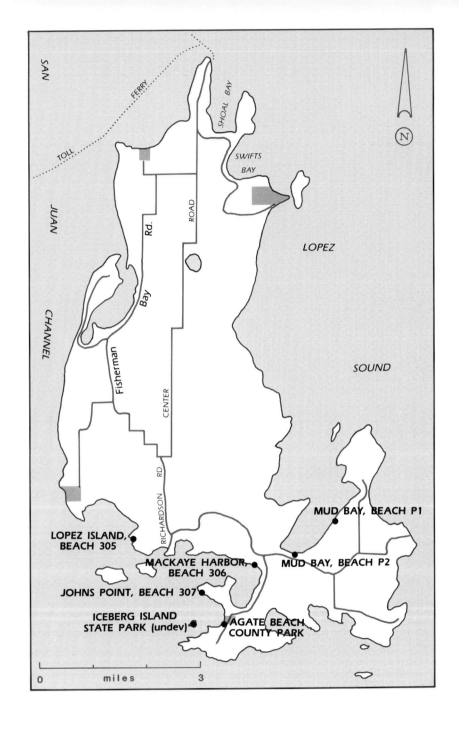

SAN JUAN COUNTY
LOPEZ ISLAND

	ACRES	CAMP UNITS	PICNIC UNITS	RESTROOMS	FIREPITS	SWIMMING BEACH	BOAT LAUNCH (lanes)	BOAT MOORAGE (slips/buoys)	PUBLIC PIER	DRINKING WATER	VIEWPOINT	SHORELINE LENGTH (feet)	ROCK BEACH	SAND BEACH	GRAVEL BEACH	MUD BEACH	SAND DUNES	TIDE POOLS	WETLANDS	BLUFFS
AGATE BEACH COUNTY PARK	2.0		1									300		•	•					•
ICEBERG ISLAND STATE PARK (undev)	3.6											1,599	•		•					•
JOHNS POINT, BEACH 307	NA											3,751	•		•					•
LOPEZ ISLAND, BEACH 305	NA											4,055			•					
MACKAYE HARBOR, BEACH 306	NA											2,580	•		•					•
MUD BAY, BEACH P1	NA											5,986		•	•					
MUD BAY, BEACH P2	NA											5,372		•		•				

AGATE BEACH COUNTY PARK
Located on Lopez Island. A small county park with limited development and facilities. No fires or camping.

ICEBERG ISLAND STATE PARK (undev)
Located 0.25 mile off the southwest side of Lopez island. Good anchorage on the east side of the island. An undeveloped area with no facilities.

JOHNS POINT, BEACH 307
Johns Point is located just north of Outer Bay on Lopez Island. Access is by boat only. The steep rocky headlands with rocky reefs and gravel pocket beaches are typical of this area. Only the tidelands are public.

LOPEZ ISLAND, BEACH 305
Located next to Davis Bay on the western side of Lopez Island. Access is by boat only. The beach has a rocky bluff with pea gravel pockets. The site also includes a group of exposed reefs in the cove. The offshore islands and reefs are nature preserves not open to the public. Only the tidelands are public.

MACKAYE HARBOR, BEACH 306
Located towards the south end of Lopez Island in Mackaye Harbor. Access is by boat only. The area has steep rocky headlands with rocky reefs and gravel pocket beaches. Only the tidelands are public.

MUD BAY, BEACH P1
Located on the east side of Mud Bay on Lopez Island. No recreation facilities. Access is by boat only. Extensive mud at low tide. Only the tidelands are public.

MUD BAY, BEACH P2
Located in Mud Bay on Lopez Island. No recreation facilities available. Access is by boat only. Extensive mud and fine sand at low tide. Only the tidelands are public.

SAN JUAN COUNTY
LOPEZ ISLAND

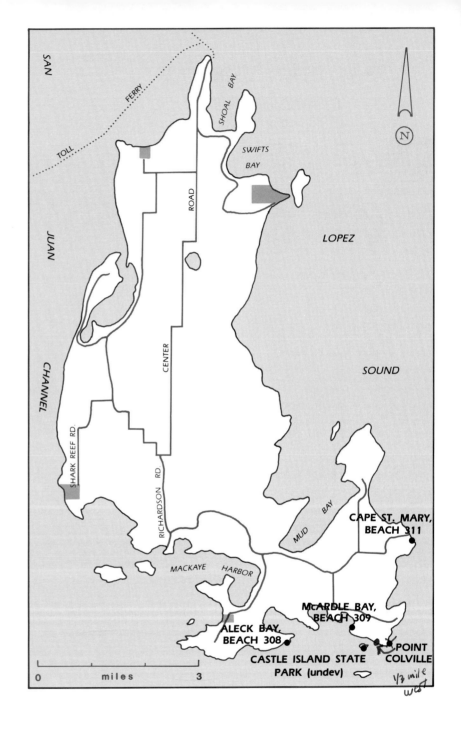

108

SAN JUAN COUNTY
LOPEZ ISLAND

	ACRES	CAMP UNITS	PICNIC UNITS	RESTROOMS	FIREPITS	SWIMMING BEACH	BOAT LAUNCH (lanes)	BOAT MOORAGE (slips/buoys)	PUBLIC PIER	DRINKING WATER	VIEWPOINT	SHORELINE LENGTH (feet)	ROCK BEACH	SAND BEACH	GRAVEL BEACH	MUD BEACH	SAND DUNES	TIDE POOLS	WETLANDS	BLUFFS
ALECK BAY, BEACH 308	NA											7,132	•		•					•
CAPE ST. MARY, BEACH 311	NA											11,557	•		•					•
CASTLE ISLAND STATE PARK (undev)	2.0											1,100								•
McARDLE BAY, BEACH 309	NA											10,932	•		•					•
POINT COLVILLE	60.0											2,000	•		•			•		•

ALECK BAY, BEACH 308
Located on the south side of Lopez Island. Access is by boat only. The headlands are steep and rocky. Only the tidelands are public.

CAPE ST. MARY, BEACH 311
Located on the east shore of Lopez Island. Only the tidelands are public. Access is by boat only. Rocky reef and headlands with cobble and pea gravel pocket beaches.

CASTLE ISLAND STATE PARK (undev)
Located just south of Lopez Island. Undeveloped rocky island. Bird habitat; please avoid disturbing the birds.

McARDLE BAY, BEACH 309
Located near the southeastern tip of Lopez Island, this beach runs from just north of McArdle Bay, south to the point adjacent to Castle Island. Composition is rocky reefs and headlands with cobble and pea gravel pocket beaches. Only the tidelands are public.

POINT COLVILLE
Located on the southeast end of Lopez Island, at the end of Watmough Bay Road. County road ends at property. Private residences on both sides. Park on the road and walk through the woods to the bluffs. Good views and nice picnic sites. Small tidepools at low tide. Seabirds nest on offshore rocks. Old growth forest. Tidelands are state owned.

Photo by Brian Walsh

SAN JUAN COUNTY
BLAKELY/DECATUR ISL.

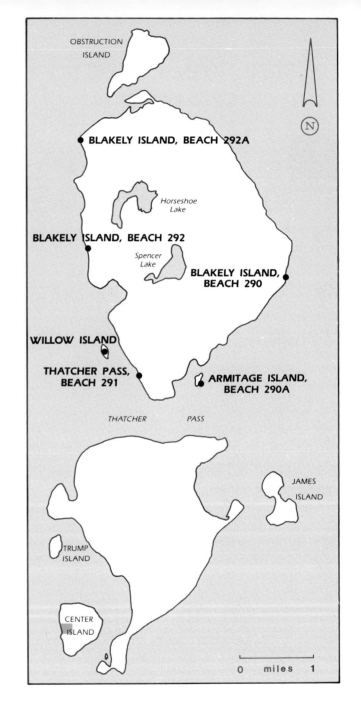

SAN JUAN COUNTY
BLAKELY/DECATUR ISL.

	ACRES	CAMP UNITS	PICNIC UNITS	RESTROOMS	FIREPITS	SWIMMING BEACH	BOAT LAUNCH (lanes)	BOAT MOORAGE (slips/buoys)	PUBLIC PIER	DRINKING WATER	VIEWPOINT	SHORELINE LENGTH (feet)	ROCK BEACH	SAND BEACH	GRAVEL BEACH	MUD BEACH	SAND DUNES	TIDE POOLS	WETLANDS	BLUFFS
ARMITAGE ISLAND, BEACH 290A	NA											2,284	•							
BLAKELY ISLAND, BEACH 290	NA											27,935			•					
BLAKELY ISLAND, BEACH 292	NA											7,649			•					•
BLAKELY ISLAND, BEACH 292A	NA											2,121			•					
THATCHER PASS, BEACH 291	NA											11,283	•	•						•
WILLOW ISLAND	9.0											2,200	•		•					

ARMITAGE ISLAND, BEACH 290A
Located off the southeast shore of Blakely Island. This beach includes the entire shoreline of Armitage Island. Access is by boat only. Only the tidelands are public.

BLAKELY ISLAND, BEACH 290
Includes most of the eastern shoreline of Blakely Island. Access is by boat only. The uplands are private. The beaches are typically narrow and steep, with a mixture of gravel and boulders. Better sandy-gravel beaches are at the south end.

BLAKELY ISLAND, BEACH 292
Located on the center-west shoreline of Blakely Island. Access is by boat only. The uplands are private. A narrow gravel beach with rocky headlands.

BLAKELY ISLAND, BEACH 292A
Located on the northwest corner of Blakely Island. Access is by boat only. The uplands are private. A gravel beach.

THATCHER PASS, BEACH 291
Located along the southern tip of Blakely Island at Thatcher Pass. Access is by boat only. Area consists of rocky headlands with sand or cobble patches and a pea gravel pocket beach on the south end. Only the tidelands are public.

WILLOW ISLAND
A small island 0.25 mile off the southwest shore of Blakely Island. This island is part of the San Juan Islands National Wildlife Refuge and is CLOSED TO THE PUBLIC to avoid disturbance of seabird nesting. Observe birds from boats at least 100 yards offshore.

SAN JUAN COUNTY
BLAKELY/DECATUR ISL.

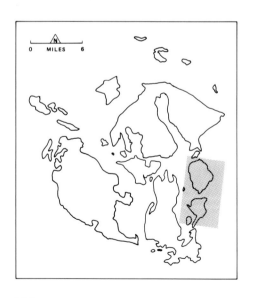
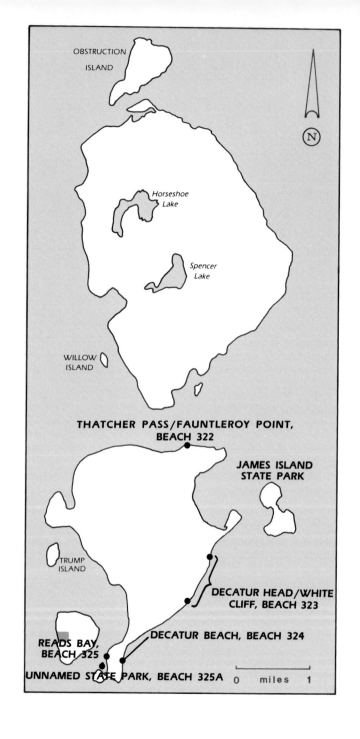

SAN JUAN COUNTY
BLAKELY/DECATUR ISL.

	ACRES	CAMP UNITS	PICNIC UNITS	RESTROOMS	FIREPITS	SWIMMING BEACH	BOAT LAUNCH (lanes)	BOAT MOORAGE (slips/buoys)	PUBLIC PIER	DRINKING WATER	VIEWPOINT	SHORELINE LENGTH (feet)	ROCK BEACH	SAND BEACH	GRAVEL BEACH	MUD BEACH	SAND DUNES	TIDE POOLS	WETLANDS	BLUFFS
DECATUR BEACH, BEACH 324	NA											1,650	•	•						
DECATUR HEAD/WHITE CLIFF, BEACH 323	NA											12,516	•	•						
JAMES ISLAND STATE PARK	113.7	13	1	•	•			10	•		•	12,340		•				•		•
READS BAY, BEACH 325	NA											1,986	•	•						
THATCHER PASS/FAUNTLEROY POINT, BEACH 322	NA											16,219	•	•	•					
UNNAMED STATE PARK, BEACH 325A	NA											1,600								

DECATUR BEACH, BEACH 324
The beach is on the southern peninsular point of Decatur Island which is located east of the mid-section of Lopez Island. Access is by boat only. Beach composition is gravel and sand with rocky bedlands. Only the tidelands are public.

DECATUR HEAD/WHITE CLIFF, BEACH 323
Located on the eastern side of Decatur Island just south of Decatur Head. Access is by boat only. The beach is very exposed and composed of cobble and sand. Only the tidelands are public.

JAMES ISLAND STATE PARK
Boating access only. Located on the east side of Decatur Island. There is a moorage float and dock with 2 mooring buoys and room on the float for about 5 boats. Overnight moorage and camping fee. There are 3 separate camping areas, each with its own crescent beach. The island is rocky with firs and madronas. Eagles live on the island.

READS BAY, BEACH 325
Located close to the southern tip of Decatur Island, bordering on Reads Bay. Access is by boat only. The area is gravel and sand with rocky bedlands. Only the tidelands are public.

THATCHER PASS/FAUNTLEROY POINT, BEACH 322
Located along the northern edge of Decatur Island, bordering on Thatcher Pass. Access is by boat only. The upper beach consists of boulders, whereas the lower beach is pea gravel and coarse sand grading into a rocky ledge with sand pockets at Fauntleroy Point. Only the tidelands are public.

UNNAMED STATE PARK, BEACH 325A
Located at the southern tip of Decatur Island. Undeveloped island with no recreation facilities, managed by the Washington State Parks and Recreation Commission.

SAN JUAN COUNTY
BLAKELY/DECATUR ISL.

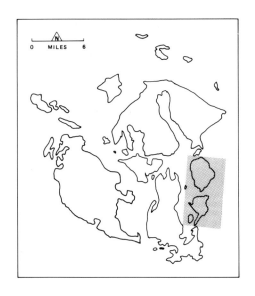
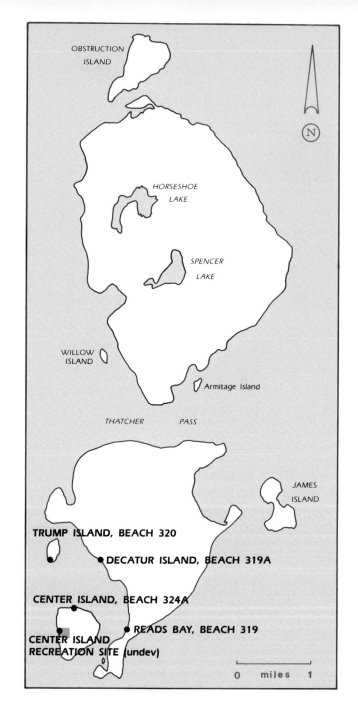

SAN JUAN COUNTY
BLAKELY/DECATUR ISL.

	ACRES	CAMP UNITS	PICNIC UNITS	RESTROOMS	FIREPITS	SWIMMING BEACH	BOAT LAUNCH (lanes)	BOAT MOORAGE (slips/buoys)	PUBLIC PIER	DRINKING WATER	VIEWPOINT	SHORELINE LENGTH (feet)	ROCK BEACH	SAND BEACH	GRAVEL BEACH	MUD BEACH	SAND DUNES	TIDE POOLS	WETLANDS	BLUFFS
CENTER ISLAND RECREATION SITE (undev)	4.2											525	•							•
CENTER ISLAND, BEACH 324A	NA											11,322	•	•						•
DECATUR ISLAND, BEACH 319A	NA											2,508			•					
READS BAY, BEACH 319	NA											2,211			•					
TRUMP ISLAND, BEACH 320	NA											3,599	•		•					•

CENTER ISLAND RECREATION SITE (undev)
A small area on the west side of Center Island which is located between Lopez and Decatur islands. Undeveloped with no facilities and generally not suitable for public use.

CENTER ISLAND, BEACH 324A
Located in Lopez Sound between Decatur and Lopez islands. Access is by boat only. All of the island's tidelands are public. The uplands are private. Most of the island has rocky headlands with pocket beaches. There is an extensive sand and gravel beach on the south side.

DECATUR ISLAND, BEACH 319A
Located on the western edge of Decatur Island, across from Lopez Island. Access is by boat only. The upper and lower beaches have different compositions. The upper beach is pea gravel and sand whereas the lower beach is cobble with some scattered boulders. Only the tidelands are public.

READS BAY, BEACH 319
Borders Reads Bay off Decatur Island, located opposite the eastern edge of Lopez Island. Access is by boat only. The beach is gravel. Only the tidelands are public.

TRUMP ISLAND, BEACH 320
Located off the western side of Decatur Island in Lopez Sound. Access is by boat only. The area is steep and rocky with reefs and sand or gravel pockets. Only the tidelands are public.

SAN JUAN COUNTY
SHAW ISLAND

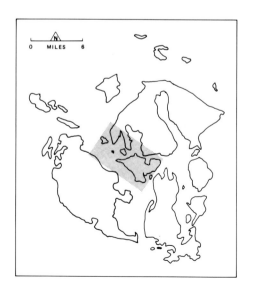

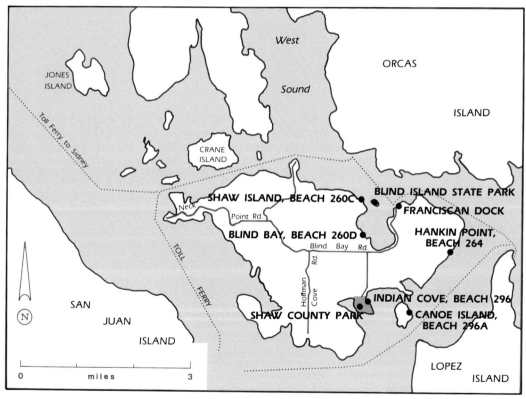

SAN JUAN COUNTY
SHAW ISLAND

	ACRES	CAMP UNITS	PICNIC UNITS	RESTROOMS	FIREPITS	SWIMMING BEACH	BOAT LAUNCH (lanes)	BOAT MOORAGE (slips/buoys)	PUBLIC PIER	DRINKING WATER	VIEWPOINT	SHORELINE LENGTH (feet)	ROCK BEACH	SAND BEACH	GRAVEL BEACH	MUD BEACH	SAND DUNES	TIDE POOLS	WETLANDS	BLUFFS
BLIND BAY, BEACH 260D	NA											2,171	•	•	•					
BLIND ISLAND STATE PARK	1.8	4	•					4				100	•					•		•
CANOE ISLAND, BEACH 296A	NA											5,196	•		•					
FRANCISCAN DOCK	NA			•				20				NA								
HANKIN POINT, BEACH 264	NA											11,344	•		•					
INDIAN COVE, BEACH 296	NA											3,249			•				•	
SHAW COUNTY PARK	65.0	8	10	•	•		1					5,800		•	•					
SHAW ISLAND, BEACH 260C	NA											1,927	•		•					

BLIND BAY, BEACH 260D
Located in Blind Bay on Shaw Island. Boat access only. The uplands are private. The eastern two-thirds is rocky with pocket beaches. The western one-third is a cobble beach with an extensive sandflat at low tide.

BLIND ISLAND STATE PARK
Located in Blind Bay on Shaw Island. A rocky shore with a small beach which can be difficult to land on. Boating access only. No drinking water, but some campsites and a pit toilet.

CANOE ISLAND, BEACH 296A
Located in Upright Channel between Lopez and Shaw islands. Includes all the tidelands of the island. The uplands are private. Rocky headlands with pocket beaches.

FRANCISCAN DOCK
Located next to the Shaw Island Ferry Terminal. Moorage fee for overnight use. Nearby general store, fuel and phone.

HANKIN POINT, BEACH 264
Located along the eastern edge of Shaw Island, just south of Orcas Island. Access is by boat only. These rocky beaches have patches of gravel. Only the tidelands are public.

INDIAN COVE, BEACH 296
Located in Indian Cove on the southeastern side of Shaw Island. The beach is accessible through Shaw Island County Park and has an extensive gravel section with rocky headlands. Only the tidelands are public.

SHAW COUNTY PARK
Located on Shaw Island. From the ferry take Blind Bay Road to Squaw Bay Road, and then take Indian Cove Road. The boat launch is good at high tide only. A grassy picnic area with a quiet sandy beach.

SHAW ISLAND, BEACH 260C
Located on the northern side of Shaw Island, directly across from Blind Island. Access is by boat only. The area is rocky with small pocket beaches. Only the tidelands are public.

SAN JUAN COUNTY
SHAW ISLAND

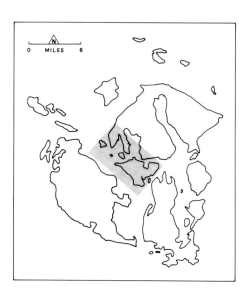

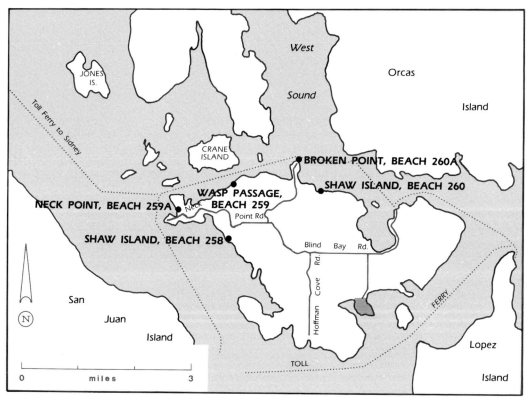

SAN JUAN COUNTY
SHAW ISLAND

	ACRES	CAMP UNITS	PICNIC UNITS	RESTROOMS	FIREPITS	SWIMMING BEACH	BOAT LAUNCH (lanes)	BOAT MOORAGE (slips/buoys)	PUBLIC PIER	DRINKING WATER	VIEWPOINT	SHORELINE LENGTH (feet)	ROCK BEACH	SAND BEACH	GRAVEL BEACH	MUD BEACH	SAND DUNES	TIDE POOLS	WETLANDS	BLUFFS
BROKEN POINT, BEACH 260A	NA											4,039	•	•	•					
NECK POINT, BEACH 259A	NA											12,021	•	•						•
SHAW ISLAND, BEACH 258	NA											1,607	•	•						•
SHAW ISLAND, BEACH 260	NA											4,620	•	•	•					
WASP PASSAGE, BEACH 259	NA											7,403	•		•					

BROKEN POINT, BEACH 260A
Located on the north side of Shaw Island, this beach includes all the tidelands of Broken Point. The uplands are private. The beach is a shallow layer of sand and gravel over a clay shelf.

NECK POINT, BEACH 259A
Located on the west side of Shaw Island. Access is by boat only. Rocky headlands are dotted with sand and gravel pocket beaches. Only the tidelands are public.

SHAW ISLAND, BEACH 258
Located on the western edge of Shaw Island, bordering San Juan Channel. Access is by boat only. The area consists of rocky headlands with sand and gravel pocket beaches. Only the tidelands are public.

SHAW ISLAND, BEACH 260
Located on the north side of Shaw Island, just east of Broken Point. Access is by boat only. The beach is narrow and rocky with sand and gravel patches or pockets. Only the tidelands are public.

WASP PASSAGE, BEACH 259
Located on the northern edge of Shaw Island, across Wasp Passage from Crane Island. Access is by boat only. The beach is narrow and rocky with sand and gravel patches or pockets. Only the tidelands are public.

SAN JUAN COUNTY
SHAW ISLAND

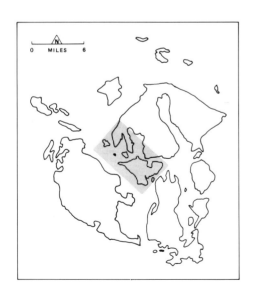

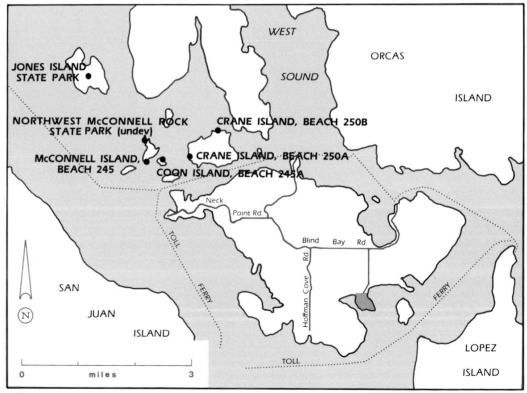

SAN JUAN COUNTY
SHAW ISLAND

	ACRES	CAMP UNITS	PICNIC UNITS	RESTROOMS	FIREPITS	SWIMMING BEACH	BOAT LAUNCH (lanes)	BOAT MOORAGE (slips/buoys)	PUBLIC PIER	DRINKING WATER	VIEWPOINT	SHORELINE LENGTH (feet)	ROCK BEACH	SAND BEACH	GRAVEL BEACH	MUD BEACH	SAND DUNES	TIDE POOLS	WETLANDS	BLUFFS
COON ISLAND, BEACH 245A	NA											1,891	•	•						•
CRANE ISLAND, BEACH 250A	NA											3,933	•	•						•
CRANE ISLAND, BEACH 250B	NA											1,452	•	•	•					
JONES ISLAND STATE PARK	179.0	21	•	•				17		•	•	25,000		•				•		
McCONNELL ISLAND, BEACH 245	NA											2,066	•	•						•
NORTHWEST McCONNELL ROCK STATE PARK (undev)	2.5											2,000	•					•		

COON ISLAND, BEACH 245A
Located off the southwest point of Orcas Island, just southeast of the small island of McConnell. Access is by boat only. Rocky headlands and reefs with nice sand and gravel pocket beaches. Only the tidelands are public.

CRANE ISLAND, BEACH 250A
Located south of Orcas Island's Deer Harbor which is on the southwest tip of Orcas. Access is by boat only. The island contains rocky headlands and reefs. Only the tidelands are public.

CRANE ISLAND, BEACH 250B
Located along the northern portion of Crane Island near Pole Pass which runs between Orcas and Crane islands. Access is by boat only. Narrow, rocky beaches with sand and gravel patches. Only the tidelands are public.

JONES ISLAND STATE PARK
Located southwest of Orcas Island. Jones Island can be reached by boat only. The entire island is a park. There are several protected harbors with interconnecting trails. Campsites at both the north and south coves have tables and stoves with nearby water faucets, garbage cans and pit toilets. The island has open grassy areas, though most of the island is forested. Creatures on the island include a number of deer, eagles, ravens and seagulls, plus several wild mink.

McCONNELL ISLAND, BEACH 245
A small island located off the southwestern tip of Orcas Island. Access is by boat only. The area consists of rocky headlands and reefs with sand and gravel pocket beaches. Only the southern half of the island's tidelands are public.

NORTHWEST McCONNELL ROCK STATE PARK (undev)
Located at the northwest corner of McConnell Island which is northwest of Shaw Island. Access is by boat only. Small undeveloped island. No moorage slips or buoys. It is an island at high tide, a spit at low tide.

ISLAND COUNTY

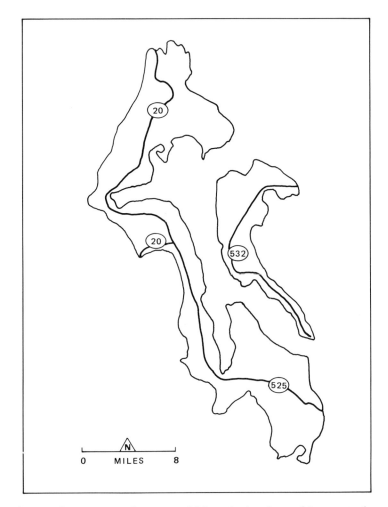

Island County consists of two islands, Camano and Whidbey, which together have 240 miles of mostly undeveloped shoreline. Whidbey Island is sixty miles long, making it the second longest island in the contiguous United States. The west side of Whidbey is exposed to wind and wave action from the Straits of Juan de Fuca and is characterized by high eroding bluffs which feed sandy beaches. The shore along the protected east side of Whidbey Island and the shore of Camano Island are mostly stable, with gravel and mud beaches.

Both islands have substantial residential and second home development. Camano Island is mostly forested while Whidbey has a lot of agricultural land. Many of Whidbey's natural prairies have been preserved, and provide sweeping views of sea and landscapes.

The north half of Whidbey Island is in the Olympic rain shadow which makes it dry enough for cactus to grow in some places. While cloud cover is similar to the rest of Puget Sound, rainfall averages as little as seventeen inches per year at Coupville. Precipitation nearly doubles at the south end of the island to match Seattle's rainfall of about thirty inches per year.

Deception Pass State Park, at the north end of Whidbey Island, is one of the main recreational attractions in western Washington. The 1,800 acre park has a wide variety of marine environments, plus several historical points of interest. A high bridge in the park connecting Whidbey Island to Fidalgo Island offers an excellent view of rushing tidal currents and whirlpools below. Further to the south is Joseph Whidbey State Park which offers day use facilities and exposure to West Beach and the Strait of Juan de Fuca.

On the mid portion of Whidbey Island is Ebey's Landing National Historic Reserve. Several old structures—old log block houses of the 1850's, Ft. Casey, Ft. Ebey, and Ebey's Landing—have been renovated here. The old port town of Coupeville, which is in the historic reserve, is a town that has retained its Victorian nineteenth century waterfront identity.

Oak Harbor, north of Coupeville, is the largest town in the county and has a major community park abutting the shoreline. Oak Harbor is surrounded by the Whidbey Naval Air Station and the majority of its population is military personnel. To the south of Coupeville is the Keystone Ferry Terminal where you can catch a ride across to Port Townsend and then drive to the Olympic Peninsula.

Southern Whidbey Island is accessible by the ferry landing at Clinton and by road from the north. Clinton is a small coastal community having a public pier and nearby public beaches.

ISLAND COUNTY
WHIDBEY ISL. NORTH

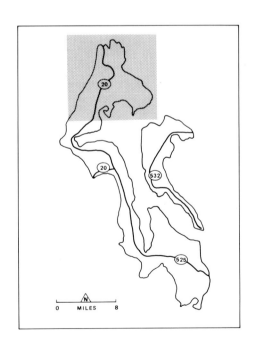
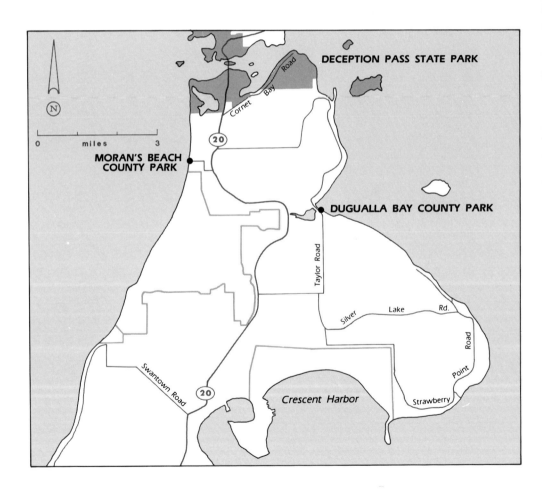

ISLAND COUNTY
WHIDBEY ISL. NORTH

	ACRES	CAMP UNITS	PICNIC UNITS	RESTROOMS	FIREPITS	SWIMMING BEACH	BOAT LAUNCH (lanes)	BOAT MOORAGE (slips/buoys)	PUBLIC PIER	DRINKING WATER	VIEWPOINT	SHORELINE LENGTH (feet)	ROCK BEACH	SAND BEACH	GRAVEL BEACH	MUD BEACH	SAND DUNES	TIDE POOLS	WETLANDS	BLUFFS
DECEPTION PASS STATE PARK	2,474.3	254	245	●	●	●	7	64		●	●	77,000	●	●	●			●	●	●
DUGUALLA BAY COUNTY PARK	NA											100			●					
MORAN'S BEACH COUNTY PARK	NA											60		●						

DECEPTION PASS STATE PARK
Located 18 miles west of Mt. Vernon on Highway 20. This is one of the major and most heavily used parks in the state. It offers a variety of facilities and opportunities for recreational use. The park encompasses 8 islands. Drive past the Cornet Bay boat launch on a dead end road to find an unspoiled beach. Freshwater swimming is possible in Cranberry Lake. Much of the park is old-growth forest. Other features of the park include an environmental center, moorage and docks, an amphitheater, and 8.5 miles of hiking trails. Showers are available at Cornet Bay.

DUGUALLA BAY COUNTY PARK
Located on the Dugualla Bay Road on the dike. A wide shoulder provides limited parking for access to the dike. Not a very usable area. No facilities.

MORAN'S BEACH COUNTY PARK
Located at the end of Powell Road adjacent to the north boundary of the Whidbey Island Naval Air Station. Beaches to the south and to the north are not publicly owned according to signs, but apparently public use is allowed.

ISLAND COUNTY
WHIDBEY ISL. NORTH

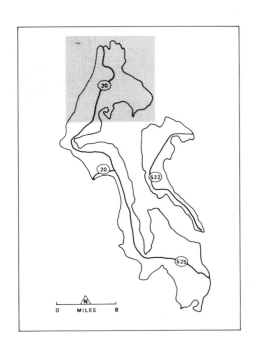
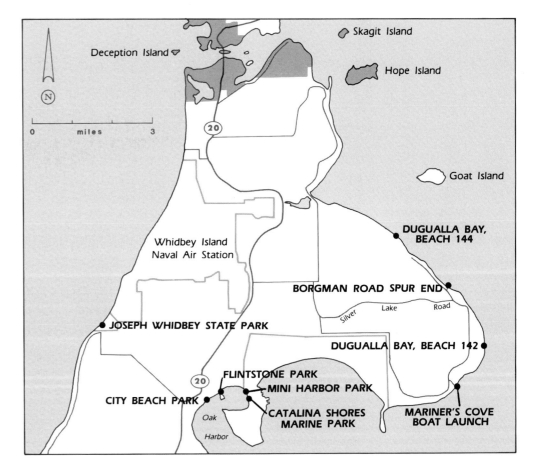

ISLAND COUNTY
WHIDBEY ISL. NORTH

PUBLIC SHORE

	ACRES	CAMP UNITS	PICNIC UNITS	RESTROOMS	FIREPITS	SWIMMING BEACH	BOAT LAUNCH (lanes)	BOAT MOORAGE (slips/buoys)	PUBLIC PIER	DRINKING WATER	VIEWPOINT	SHORELINE LENGTH (feet)	ROCK BEACH	SAND BEACH	GRAVEL BEACH	MUD BEACH	SAND DUNES	TIDE POOLS	WETLANDS	BLUFFS
CATALINA SHORES MARINE PARK	7.3			●			8	316	●			1,000								
CITY BEACH PARK	30.0	55	15	●	●	●			●	●		2,100	●		●		●			
DUGUALLA BAY, BEACH 142	NA											4,800			●					
DUGUALLA BAY, BEACH 144	NA											4,800			●					
FLINTSTONE PARK	NA		2						●			500	●							
JOSEPH WHIDBEY STATE PARK	112.0		5	●	●							3,100		●						
MARINER'S COVE BOAT LAUNCH	0.2						1			●		40		●						
MINI HARBOR PARK	2.0		2	●								500								
ROAD END OF SPUR ROAD OFF BORGMAN ROAD	NA											40		●						

CATALINA SHORES MARINE PARK
Located on the east end of the town of Oak Harbor. Mostly small boat moorage and boat launch. A small day use area is located next door at Mini Harbor Park.

CITY BEACH PARK
Located on the waterfront of Oak Harbor on Whidbey Island. Beach is an extensive mudflat at low tide. Park has a large lawn and is well maintained. The park provides a wide variety of activities.

DUGUALLA BAY, BEACH 142
Located on the northern portion of Whidbey Island, east of Oak Harbor. Access is by boat only. All surrounding property is privately owned and there is no trespassing. Only the tidelands are public.

DUGUALLA BAY, BEACH 144
Located on the eastern side of Whidbey Island in the northern section. Access is available by boat, by car or by hiking through the state land. No facilities for recreational use.

FLINTSTONE PARK
Located on the waterfront of Oak Harbor on Whidbey Island. A pier and float provide boating access. A new park developed according to the theme of the comic strip.

JOSEPH WHIDBEY STATE PARK
Located on Crosby Road on the south side of Whidbey Island Naval Air Station. Recently acquired and held for future development. Limited facilities.

MARINER'S COVE BOAT LAUNCH
Located within the Mariners Cove subdivision near the end of North Beach Drive which is off Strawberry Road. No signs. View of Mt. Baker and Glacier Peak. No facilities other than the launch lane and a small parking area.

MINI HARBOR PARK
This park is adjacent to Catalina Shores Marine Park. The beach is small and not very usable.

ROAD END OF SPUR ROAD OFF BORGMAN ROAD
This street end provides access to extensive public tidelands. A low bank has private lots on both sides. Could launch a boat here. Residential on both sides.

ISLAND COUNTY
WHIDBEY ISL. CENTER

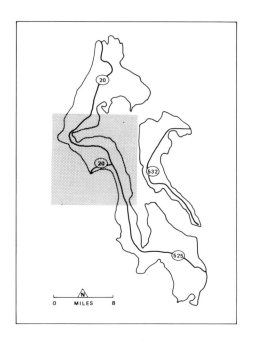

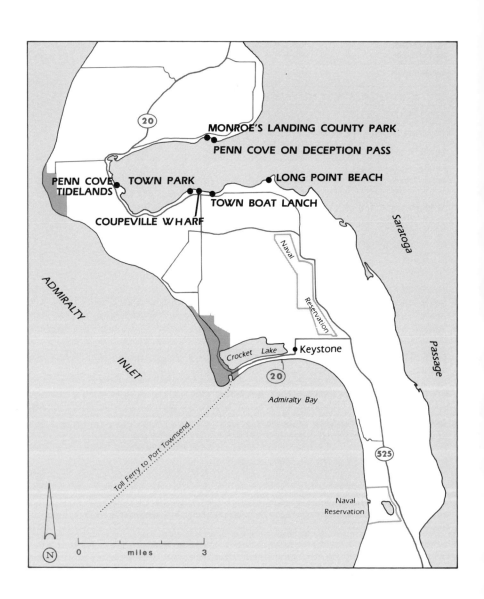

ISLAND COUNTY
WHIDBEY ISL. CENTER

	ACRES	CAMP UNITS	PICNIC UNITS	RESTROOMS	FIREPITS	SWIMMING BEACH	BOAT LAUNCH (lanes)	BOAT MOORAGE (slips/buoys)	PUBLIC PIER	DRINKING WATER	VIEWPOINT	SHORELINE LENGTH (feet)	ROCK BEACH	SAND BEACH	GRAVEL BEACH	MUD BEACH	SAND DUNES	TIDE POOLS	WETLANDS	BLUFFS
COUPEVILLE WHARF	NA						8	•	•			NA		•					•	
LONG POINT BEACH	NA							•				60			•					
MONROE'S LANDING COUNTY PARK	0.1						1					60	•	•						
PENN COVE ON DECEPTION PASS	0.4											NA	•	•						
PENN COVE TIDELANDS	25.0											15,000								
TOWN BOAT LAUNCH	1.0						2			•		750								
TOWN PARK	3.0	13	•	•							•	450		•						•

COUPEVILLE WHARF
Located at the corner of Front and N.W. Alexander. This historic wharf extends into Penn Cove. Interpretive exhibits about historic Coupeville and the Ebey's Landing historic district. There is an overnight moorage float at the end of the wharf. Fee charged.

LONG POINT BEACH
Located at the west end of Marine Drive. Unimproved road end access with a low bank waterfront. A favorite smelt raking area. Scenic vista of the cliffs on Northern Whidbey Island.

MONROE'S LANDING COUNTY PARK
Located on Penn Cove Road at the intersection of Monroe's Landing Road. Provides access to extensive public tidelands but has no facilities except for the boat launch ramp.

PENN COVE ON DECEPTION PASS
Located at the end of Monroe's Landing Road. Tidelands owned by the State Department of Game. *by end of 1987. Funded improved public access will include parking and pit toilets.*

PENN COVE TIDELANDS
The northwest end of Penn Cove historically supported extensive sport harvest of hard shell clams. Their harvest has all but stopped due to a lack of parking and rigid enforcement of trespass laws. The Department of Fisheries does own access opposite Kennedy's Lagoon but the boundaries are controversial and must be surveyed.

I goofed

TOWN BOAT LAUNCH
Located next to the city sewage treatment plant on N.E. 9th Street in Coupeville. The beach is mostly bulkheaded. A trailer holding tank dump station is located on the site.

TOWN PARK
Located at the top of the hill on the west end of town on N.W. Coveland Street. The park has an expansive lawn with big shade trees. Not much beach at high tide. View across Penn Cove of north Whidbey Island.

ISLAND COUNTY
WHIDBEY ISL. CENTER

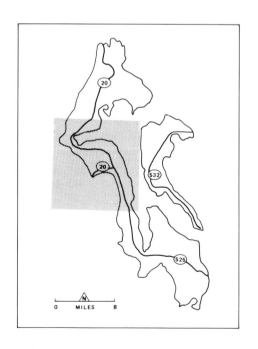
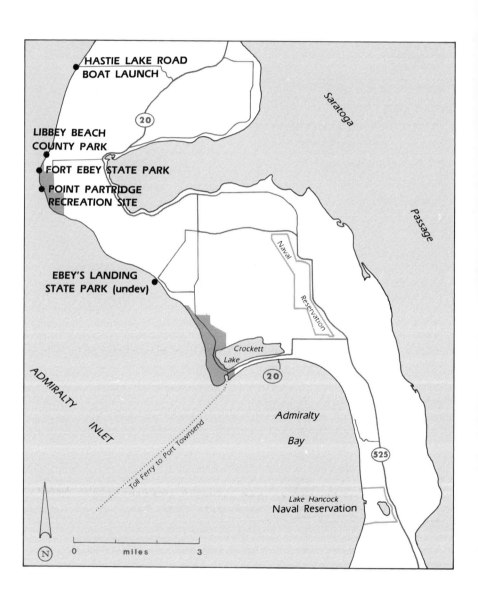

ISLAND COUNTY
WHIDBEY ISL. CENTER

PUBLIC SHORE

	ACRES	CAMP UNITS	PICNIC UNITS	RESTROOMS	FIREPITS	SWIMMING BEACH	BOAT LAUNCH (lanes)	BOAT MOORAGE (slips/buoys)	PUBLIC PIER	DRINKING WATER	VIEWPOINT	SHORELINE LENGTH (feet)	ROCK BEACH	SAND BEACH	GRAVEL BEACH	MUD BEACH	SAND DUNES	TIDE POOLS	WETLANDS	BLUFFS
EBEY'S LANDING STATE PARK (undev)	22.8										•	2,720		•						
FORT EBEY STATE PARK	228.2	50	20	•	•						•	9,000	•	•				•		•
HASTIE LAKE ROAD BOAT LAUNCH	0.8						1				•	100	•	•						
LIBBEY BEACH COUNTY PARK	3.1		5	•	•						•	300	•							•
POINT PARTRIDGE RECREATION SITE	23.0	11	4	•	•							9,000		•						•

EBEY'S LANDING STATE PARK (undev)
Located on the eastern shore of Whidbey Island where the only access is at a gravel parking lot on Hill Road at the bottom of the hill.

FORT EBEY STATE PARK
Located 5 miles out of Coupeville at Point Partridge. The camping area is in a coniferous forest on the top of the bluff. The bluff is eroding. Tidepool life amidst the rocks is abundant. The park is well kept and unspoiled. Not as heavily used as most state parks but, nonetheless, delightful to visit. This is a historical park with gun emplacements.

HASTIE LAKE ROAD BOAT LAUNCH
Located at the end of Hastie Lake Road. Provides access to extensive tidelands but because of the exposure, boat launching can be rough. A sign nearby says that the adjacent beach is private but the public is welcome to use - a first!

LIBBEY BEACH COUNTY PARK
Developed on the western exposed shore of Whidbey Island where winter storms have taken their toll. Follow Libbey Road to the end, close to Point Partridge. A boat launch ramp has been washed out and the shoreline is generally eroding. The park does provide access to extensive beaches.

POINT PARTRIDGE RECREATION SITE
Located 5 miles west of Coupeville. A primitive camping area with hiking trail to extensive public beaches. Eroding bluff extends for a mile or more where a sandy clay material is at the angle of repose.

ISLAND COUNTY
WHIDBEY ISL. CENTER

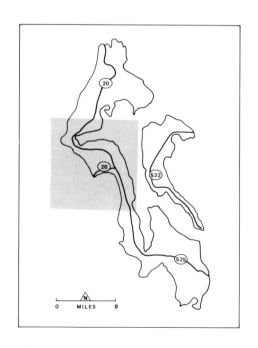
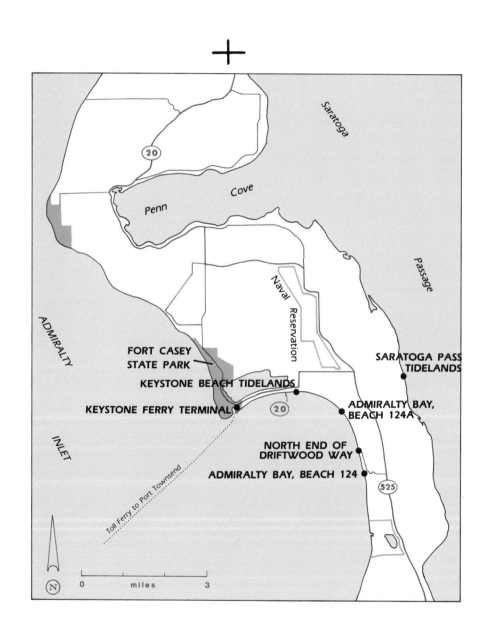

ISLAND COUNTY
WHIDBEY ISL. CENTER

PUBLIC SHORE

	ACRES	CAMP UNITS	PICNIC UNITS	RESTROOMS	FIREPITS	SWIMMING BEACH	BOAT LAUNCH (lanes)	BOAT MOORAGE (slips/buoys)	PUBLIC PIER	DRINKING WATER	VIEWPOINT	SHORELINE LENGTH (feet)	ROCK BEACH	SAND BEACH	GRAVEL BEACH	MUD BEACH	SAND DUNES	TIDE POOLS	WETLANDS	BLUFFS
ADMIRALTY BAY, BEACH 124	NA											2,400	•	•						
ADMIRALTY BAY, BEACH 124A	NA											4,200	•	•						
FORT CASEY STATE PARK	137.5	35	60	•	•		2			•	•	8,200		•					•	
KEYSTONE BEACH TIDELANDS	NA											NA		•						
KEYSTONE FERRY TERMINAL	NA		•							•		600		•						
NORTH END OF DRIFTWOOD WAY	NA											NA		•					•	
SARATOGA PASS TIDELANDS	NA											8,500	•		•					

ADMIRALTY BAY, BEACH 124
Located on Whidbey Island just south of Fort Casey State Park. Access is by boat only. The upper beach is sand and the lower beach is cobble. Only the tidelands are public.

ADMIRALTY BAY, BEACH 124A
Located just south of Fort Casey State Park. Access is by boat only. The upper beach material is sand whereas the lower beach is cobble. Only the tidelands are public.

FORT CASEY STATE PARK
Located on Admiralty Head next to the Keystone Ferry Terminal. Historical coastal artillery emplacement for the defense of Puget Sound. Camp area is crowded onto a beach area and usually jammed with recreational vehicles. Open year round. The jetty which protects the harbor also provides an underwater park for scuba divers. Removal of sea life is prohibited. A popular park which provides for a wide variety of users from kite flyers to ball players to picnickers and people just playing.

KEYSTONE BEACH TIDELANDS
Located at the end of Keystone Road. There are extensive public tidelands located here but except for the state park there is very limited public access to them. Most of the tidelands have not been surveyed or marked for public use. The uplands at this point are private.

KEYSTONE FERRY TERMINAL
Located on the west side of Whidbey Island. Provides ferry service to Port Townsend.

NORTH END OF DRIFTWOOD WAY
Located at the north end of Driftwood Way in the Ledgewood subdivision. There are some problems here with bank erosion and mass soil movement. Provides access to extensive state tidelands.

SARATOGA PASS TIDELANDS
Located on the east shore of Whidbey Island on Saratoga Passage. Only the tidelands are public. No upland access. Good clam digging at the southern end.

ISLAND COUNTY
WHIDBEY ISL. SOUTH

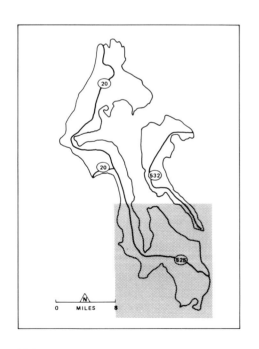
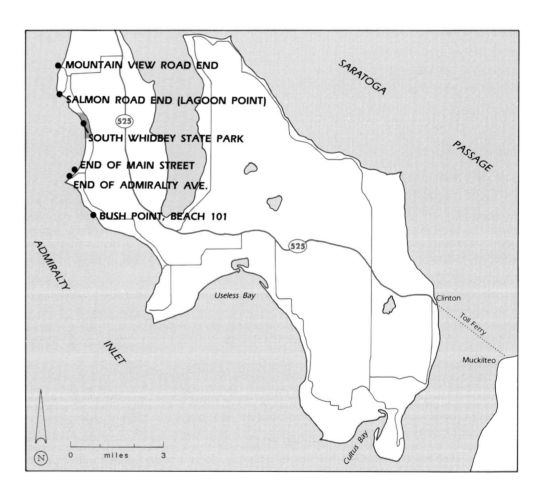

ISLAND COUNTY
WHIDBEY ISL. SOUTH

PUBLIC SHORE

	ACRES	CAMP UNITS	PICNIC UNITS	RESTROOMS	FIREPITS	SWIMMING BEACH	BOAT LAUNCH (lanes)	BOAT MOORAGE (slips/buoys)	PUBLIC PIER	DRINKING WATER	VIEWPOINT	SHORELINE LENGTH (feet)	ROCK BEACH	SAND BEACH	GRAVEL BEACH	MUD BEACH	SAND DUNES	TIDE POOLS	WETLANDS	BLUFFS
BUSH POINT, BEACH 101	NA											1,650	•	•						
END OF ADMIRALTY AVE.	NA											40	•							
END OF MAIN STREET	NA											40		•						
MOUNTAIN VIEW ROAD END	NA											40		•						
SALMON ROAD END (LAGOON POINT)	NA											20		•						
SOUTH WHIDBEY STATE PARK	85.0	70	19	•	•				•	•		4,500		•						•

BUSH POINT, BEACH 101
Located on Whidbey Island, south of South Whidbey State Park. Access is by boat only. The upper beach is made up of sand, the lower beach composed of cobble and large boulders. Only the tidelands are public.

END OF ADMIRALTY AVE.
Located on Whidbey Island at the end of Admiralty Ave. on Bush Point. Car-top boat can be launched here. Nearby Bush Point Lighthouse is not open to the public.

END OF MAIN STREET
Located at the end of Main Street at Bush Point. This unimproved street end provides a small beach, but it is not a very attractive setting. No facilities, but there is a privately operated boat hoist next door.

MOUNTAIN VIEW ROAD END
Located on the west side of Whidbey Island, this is an unimproved road end with a low bank. Possible to launch a car-top boat here.

SALMON ROAD END (LAGOON POINT)
Located at the end of Salmon Road on Lagoon Point. This is a road end access with no facilities. The county only owns the northern half of the right-of-way so the public tidelands are very narrow. Private residences on both sides.

SOUTH WHIDBEY STATE PARK
Located on Smugglers Cove Road on the west side of Whidbey Island. Large park in a coniferous forest. Trail to the beach is long and steep. Beach is very narrow and not very usable at high tide.

ISLAND COUNTY
WHIDBEY ISL. SOUTH

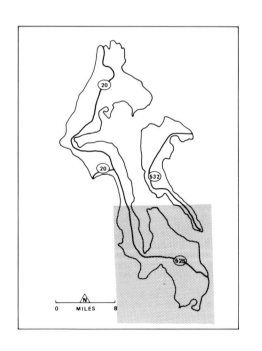
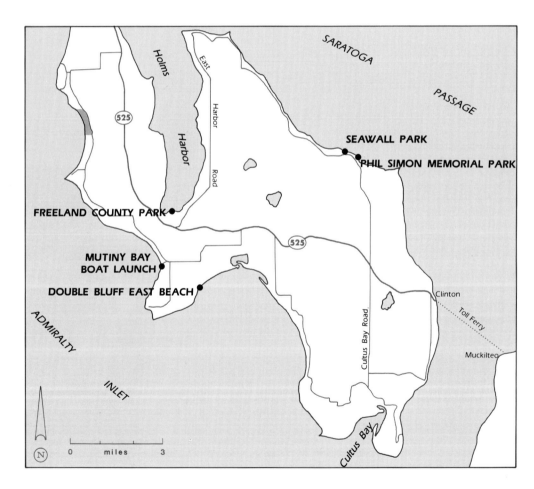

ISLAND COUNTY
WHIDBEY ISL. SOUTH

	ACRES	CAMP UNITS	PICNIC UNITS	RESTROOMS	FIREPITS	SWIMMING BEACH	BOAT LAUNCH (lanes)	BOAT MOORAGE (slips/buoys)	PUBLIC PIER	DRINKING WATER	VIEWPOINT	SHORELINE LENGTH (feet)	ROCK BEACH	SAND BEACH	GRAVEL BEACH	MUD BEACH	SAND DUNES	TIDE POOLS	WETLANDS	BLUFFS
DOUBLE BLUFF EAST BEACH	NA											40	•							
FREELAND COUNTY PARK	7.1	15	•	•	•	1				•		1,550	•							•
MUTINY BAY BOAT LAUNCH	0.5						1					60	•							
PHIL SIMON MEMORIAL PARK	NA	2	•				1		•	•	•	400	•							
SEAWALL PARK	1.0	3								•		1,000			•					•

DOUBLE BLUFF EAST BEACH
Located on Whidbey Island at the south end of Double Bluff Road. Parking area and turnaround. State owned tidelands with private uplands except for the road end. No facilities. Clamdigging during season. Beachwalking and swimming in shallow pools.

FREELAND COUNTY PARK
Located in Freeland at the south end of Holmes Harbor. A portion of this park is newly developed. Forested bluff to the west is part of the old park. Ideal place for a group picnic.

MUTINY BAY BOAT LAUNCH
Turn on Robinson Road from Mutiny Bay Road. Sandy beach on both sides of the boat launch. No facilities other than the boat launch. Parking lot is half a block away.

PHIL SIMON MEMORIAL PARK
Located in Langley at the end of a short road which turns off 1st Street. A transient moorage facility is now being redeveloped and repaired by the City of Langley. The park is currently mostly unmaintained and signs are posted about unsafe conditions during storms. The boat launch may be unusable because it has drifted in with sand.

SEAWALL PARK
Located in Langley on 1st Street. Stairs lead down to the beach. Park on the street above.

ISLAND COUNTY
WHIDBEY ISL. SOUTH

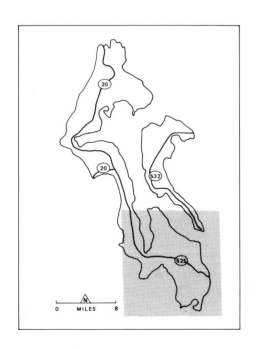
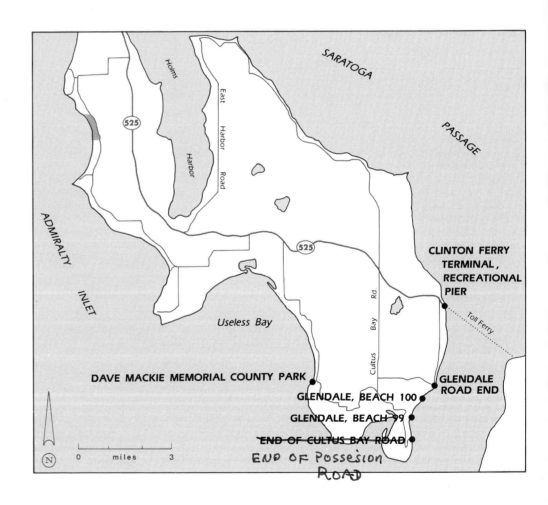

ISLAND COUNTY
WHIDBEY ISL. SOUTH

PUBLIC SHORE

	ACRES	CAMP UNITS	PICNIC UNITS	RESTROOMS	FIREPITS	SWIMMING BEACH	BOAT LAUNCH (lanes)	BOAT MOORAGE (slips/buoys)	PUBLIC PIER	DRINKING WATER	VIEWPOINT	SHORELINE LENGTH (feet)	ROCK BEACH	SAND BEACH	GRAVEL BEACH	MUD BEACH	SAND DUNES	TIDE POOLS	WETLANDS	BLUFFS
CLINTON FERRY TERMINAL	NA		•								•	NA								
CLINTON RECREATIONAL PIER	NA							2	•			NA	•							
DAVE MACKIE MEMORIAL COUNTY PARK	4.3	6	•	•	•	1				•		400	•							
END OF ~~CULTUS BAY ROAD~~ Possion Road	NA											60		•						
GLENDALE ROAD END	NA											40		•						
GLENDALE, BEACH 100	NA											2,550	•	•						
GLENDALE, BEACH 99	NA											1,160	•	•						

Possesion Beach Park (under development)
Junction Franklin Rd + Possesion Rd   680 ft waterfront

CLINTON FERRY TERMINAL
Located on the south end of Whidbey Island providing a ferry link to the mainland at Mukilteo. The facilities here are primarily intended for ferry passengers. The Clinton Recreational Pier, next door, is a public facility.

CLINTON RECREATIONAL PIER
Located on the Clinton Ferry Dock. A small float and walkway with space to moor a couple of boats. Beach nearby at the adjacent restaurant appears to be usable by the public. Parking available in the ferry commuter's lot. Moorage is only temporary and not to be used for overnight.

DAVE MACKIE MEMORIAL COUNTY PARK
Located at the end of Maxwelton Road near Indian Point. The park has a nice, low bank beach with some lawn but no shade. There are no signs.

END OF CULTUS BAY ROAD
Located at the end of Cultus Bay Road. An undeveloped site with potential for a boat launch. Parking is limited.

GLENDALE ROAD END
Located at the end of Glendale Road on the southeast shore of Whidbey Island. This is an undeveloped road end with the potential for a boat launch.

GLENDALE, BEACH 100
Located near the southern point of Whidbey Island, just south of Glendale which is south of Clinton. Access is by boat only. The upper beach is made up of sand, the lower beach made up of cobble. Only the tidelands are open to the public.

GLENDALE, BEACH 99
Located on Whidbey Island just north of Possession Point on the eastern edge of the point. Access is by boat only. The upper beach is composed of sand and the lower beach is cobble. Only the tidelands are public.

ISLAND COUNTY
CAMANO ISLAND

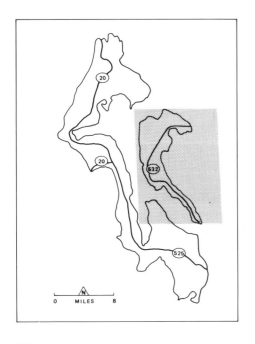

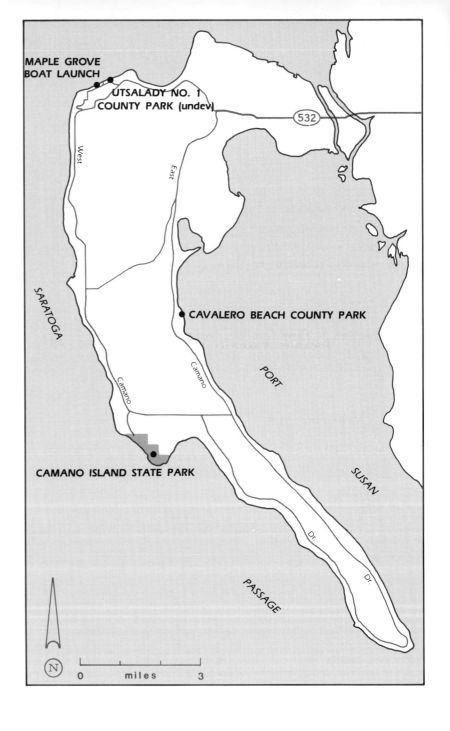

ISLAND COUNTY
CAMANO ISLAND

PUBLIC SHORE

	ACRES	CAMP UNITS	PICNIC UNITS	RESTROOMS	FIREPITS	SWIMMING BEACH	BOAT LAUNCH (lanes)	BOAT MOORAGE (slips/buoys)	PUBLIC PIER	DRINKING WATER	VIEWPOINT	SHORELINE LENGTH (feet)	ROCK BEACH	SAND BEACH	GRAVEL BEACH	MUD BEACH	SAND DUNES	TIDE POOLS	WETLANDS	BLUFFS
CAMANO ISLAND STATE PARK	134.3	87	114	•	•		2			•		6,700		•						•
CAVALERO BEACH COUNTY PARK	0.6		2				1			•		300		•	•					
MAPLE GROVE BOAT LAUNCH	0.2						1					30								
UTSALADY NO. 1 COUNTY PARK (undev)	0.1						1					40								

CAMANO ISLAND STATE PARK
Large state park located on Camano Island. Picnic area near the beach. Camping higher in the woods. No fires on the beach. Scenic view of Olympics. Clamdigging in season. High bluff behind the beach.

CAVALERO BEACH COUNTY PARK
Located on Camano Island on Cavalero Road; off East Camano Drive take Cavalero Road 0.5 mile to county park. Follow signs. Boat launch is only usable at high tide. Scenic view of Mount Baker and Port Susan. No fires.

MAPLE GROVE BOAT LAUNCH
Located on the northwest shore of Camano Island. No facilities other than a boat launch.

UTSALADY NO. 1 COUNTY PARK (undev)
Located on the north end of Camano Island. Undeveloped park with no facilities.

WASHINGTON FORTS

Much of Washington's history is recorded in the historic sites and landmarks scattered along the state's shoreline. Some of the most enjoyable and informative historic sites are the old military fortifications now owned and managed by the Washington State Parks and Recreation Commission.

Puget Sound Forts

Three major Puget Sound military forts—Fort Casey, Fort Flagler and Fort Worden—were built between 1879 and 1911. They were designed to form a "triangle of fire" to guard Admiralty Inlet, which was the only access to Puget Sound cities that could be navigated by warships at that time. All three forts have been purchased and restored, to varying degrees, by State Parks.

Fort Worden, located on Point Wilson in Jefferson County, was built on the site of an earlier fort, Fort Wilson, which was constructed in 1855 to protect the settlement of Port Townsend from Indian attacks. Fort Wilson was abandoned a year after it was built when hostilities with the Indians ended.

At the turn of the century, the fort was reactivated as Fort Worden. Buildings and gun emplacements were constructed and the headquarters of the defense of Puget Sound was transferred there. Four companies were stationed there.

Fort Worden expanded further during World War I, but when the war ended in 1919, both guns and personnel were cut back.

In 1951 military jurisdiction of the fort officially ended, and in 1957 it was purchased by the state and used as a rehabilitation institution. In 1972, Fort Worden was purchased by the

State Parks and Recreation Commission and it was developed for public use in the years following. The old Victorian houses and barracks are now available as conference facilities and vacation housing. The fort also has history walks and interpretive information.

Fort Flagler is located in Jefferson County on the north end of Marrowstone Island. Construction began on the fort in 1897 and by 1900, armaments were installed and barracks were completed. Fort Flagler was active during World War I but was placed under caretaker status in 1937.

In 1940, twenty-four new buildings were constructed and men from the Harbor Defense of Puget Sound, including two Coast Artillery regiments, moved in until 1943. From 1945 to 1954 the fort was used for training engineers and amphibious military units.

In 1955 Fort Flagler was purchased and developed as a recreation area by State Parks. The area now has natural history and historical interpretive sites and trails, and an environmental learning center.

Fort Casey, on Whidbey Island, had troops arrive and gun emplacements constructed in the late 1800s. During World War I it was used mainly for training activities and after the war the army placed the fort in caretaker status. Between 1922 and 1945 all armament was scraped, melted down and sold to help support the fort.

With the onset of World War II, Fort Casey was reactivated as a training center. Barracks were rebuilt and anti-aircraft guns were installed in the old emplacements.

After 1950 the fort became obsolete for military purposes; it was bought by State Parks in 1956.

The Admirality Head Lighthouse was built at Fort Casey in 1858. State Parks has converted the light house into an interpretive center which focuses on Fort Casey's history.

Fort Ebey, just north of Fort Casey, was established as a coast artillery fort in 1942 to supplement the artillery at Forts Casey, Flagler, and Worden. Soon after World War II the military declared the fort surplus property and the area was donated to the state in 1968. In 1980 Fort Ebey was developed as a state park and it was opened to the public a year later. Extensive natural areas around the fort make it an enjoyable place to explore the beach and woods.

Fort Ward, located on Bainbridge Island in Kitsap County, was designated as a seacoast fort in 1903 to protect the Bremerton Naval Shipyard. The fort was heavily armed, but after only a short time in operation it was put into inactive status and abandoned. However, during the World War II, a submarine net was installed across Rich Passage at Fort Ward, and radio stations which could break enemy codes were set up. In 1958, the fort was deactivated by the Army. State Parks purchased Fort Ward in 1960. Minimal development has taken place at the site and there are extensive natural areas.

Fort Hayden, in Clallam County on the Strait of Juan de Fuca, was developed during World War II by the United States in an effort to beef up coastal defenses with anti-motor torpedo boats, seacoast armament and anti-aircraft installations. By the end of the war, Fort Hayden had two heavily camouflaged, bomb-proofed gun batteries on Striped Peak and Tongue Point. The batteries covered the entrance to Puget Sound, Victoria Harbor (British Columbia), and the Canadian Naval base at Esquimalt.

In 1949, the Army declared the fort surplus and after years of discussion, Clallam County finally approved funds for purchase and turned the area into the Salt Creek Recreation Area.

Columbia River Forts

Along the Columbia River in southern Pacific County, Fort Columbia and Fort Canby were constructed to protect the entrance to the Columbia River (a British fort also was constructed across the river near Astoria, Oregon).

Fort Columbia, located near the town of Chinook, was purchased by the United States in 1867. In 1895 the War Department decided to install defensive weapons and construct housing for troops. The fort was first occupied by a garrison of troops in June of 1904. Fort Columbia continued to serve as a unit of coast defenses in the first half of the 1900s but it never fired a hostile shot.

The former enlisted men's barracks at the fort is now an interpretive center that gives visitors an idea of what life in the Coast Artillery was like. There is also a museum and an art gallery. The former Coast Artillery Hospital now operates as a youth hostel.

Fort Canby is located on Cape Disappointment where the Lewis and Clark expedition reached the ocean in 1805. The Cape Disappointment Lighthouse began operating in 1856 and is still in use, making it the oldest operating lighthouse on the West Coast.

Cape Disappointment was first armed in 1862 to protect the mouth of the Columbia River from enemies. In 1875 the area took on the name Fort Canby and was expanded for military use until the end of World War II. After that the fort was deactivated and in 1957 it was dedicated as a state park.

The Lewis and Clark interpretive center, located at the fort, has a educational multimedia program that highlights various aspects of the Lewis and Clark journey from St. Louis to Cape Disappointment, and portrays later events at the cape. The center is open seven days a week during the summer and on weekends during the winter.

Fort Canby also has other interpretive programs during the summer months and extensive natural areas to explore in all seasons.

All of these coastal fortifications, designed to protect Washington's harbors from enemy ships, became obsolete with the advent of military aircraft. The guns which were designed to shoot at targets sitting on the water could not be raised to shoot aircraft. Consequently, what were state-of-the-art military installations at the time of their construction, quickly became useless relics after the turn of the century. The anti-aircraft artillery batteries that replaced the old structures have now been made obsolete by modern missiles.

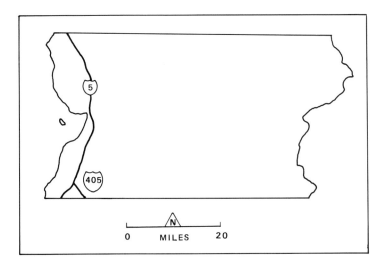

SNOHOMISH COUNTY

Snohomish County has sixty-two miles of saltwater shoreline. The shoreline consists mostly of mainland frontage along the eastern side of Puget Sound, but it also includes a large island, Gedney, commonly known as Hat Island. The county includes the major city of Everett, situated at the mouth of the Snohomish River, and the Tulalip Indian Reservation just north of Everett.

A portion of the Skagit River Estuary, an important waterfowl habitat area, is in the northern part of the county. The estuary is discussed in greater detail under Skagit County. To the south, the shoreline is highly developed, and eventually blends into the urban sprawl of the greater Seattle area.

Unfortunately, all of the shoreline south of Everett is paralleled and consumed by railroad right-of-way (the tracks move inland north of Everett). The trains move rapidly, making the crossing dangerous. In spite of the railroad, there are some spots where the public has safe access to the water. The Edmonds waterfront, Picnic Point and Mukilteo State Park are all good shoreline access areas. Nevertheless, considering the total shoreline of Snohomish County, there are really very few places where the public has access.

One especially delightful spot is the manmade jetty in Everett Harbor. An undeveloped state park within view of downtown, this area is great for bird watching and is a popular haul out area for seals. Kayakers can put in at Marine Park and paddle across to the jetty.

Snohomish County has two jumping off points for the Washington State Ferry system—Edmonds and Mukilteo. The latter provides a connection with the south end of Whidbey Island and the former provides access to the Kitsap Peninsula.

SNOHOMISH COUNTY
NORTH

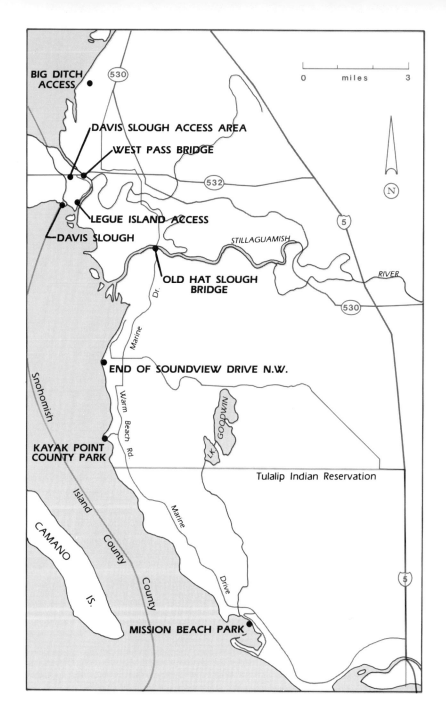
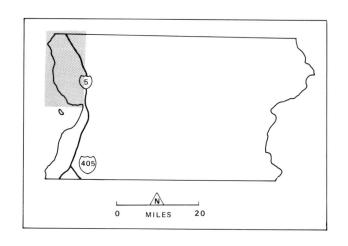

SNOHOMISH COUNTY
NORTH

	ACRES	CAMP UNITS	PICNIC UNITS	RESTROOMS	FIREPITS	SWIMMING BEACH	BOAT LAUNCH (lanes)	BOAT MOORAGE (slips/buoys)	PUBLIC PIER	DRINKING WATER	VIEWPOINT	SHORELINE LENGTH (feet)	ROCK BEACH	SAND BEACH	GRAVEL BEACH	MUD BEACH	SAND DUNES	TIDE POOLS	WETLANDS	BLUFFS
BIG DITCH ACCESS	NA											NA				•				
DAVIS SLOUGH	40.0											5,000				•			•	
DAVIS SLOUGH ACCESS AREA	NA											NA				•			•	
END OF SOUNDVIEW DRIVE N.W.	NA											60				•				
KAYAK POINT COUNTY PARK	670.0	29	70	•	•	•	1			•	•	3,300	•	•						•
LEGUE ISLAND ACCESS	123.0											NA				•				
MISSION BEACH PARK	8.0		•									400			•					
OLD HAT SLOUGH BRIDGE	NA											80	•		•		•			
WEST PASS BRIDGE	NA											50			•		•			

BIG DITCH ACCESS
Located between Milltown and Stanwood off Old Highway 99. Provides an access point to the southern part of the Skagit Habitat Management Area. See the Skagit Habitat Management Area listing under Skagit County for additional information. A State Game Department Conservation License is required to use this area.

DAVIS SLOUGH
This is an island in Davis Slough, one half mile east of Juniper Beach, Camano Island. The island is owned by the U.S. Bureau of Land Management and can be reached from the Skagit Wildlife Habitat Area.

DAVIS SLOUGH ACCESS AREA
Access is on the south side of Highway 532, 0.5 miles west of Stanwood. Davis Slough is part of the Game Department's Skagit Habitat Management Area (see write-up under Skagit County). A Conservation License is required to use this area.

END OF SOUNDVIEW DRIVE N.W.
At the south end of Sound View Drive N.W. in the community of Warm Beach, about 6 miles south of Stanwood on Marine Drive. Road end access to extensive saltwater mudflats. Parking is very limited. Park on south shoulder of road. Do not block private driveways.

KAYAK POINT COUNTY PARK
Located on Marine Drive about 8 miles south of Stanwood. Parking lot on the beach where there is a picnic area. Expansive gravelly beach has some sand. View across Port Susan to Whidbey Island. Excellent clamming. Large lawns are good for impromptu sports. Modern picnic shelter can be reserved for a fee. County owned 18-hole golf course nearby. Park hours are 6:00 a.m. until dusk.

LEGUE ISLAND ACCESS
Located at the end of Eide Road which turns south at the west end of West Pass Bridge; 0.5 miles west of Stanwood on Highway 532. Part of the Skagit Habitat Management Area. See write-up under Skagit County. A Conservation License is required to use this area.

MISSION BEACH PARK
Located on Tulalip Bay, but there are no signs. This park was developed by the Tulalip Tribe using state public recreation funding.

OLD HAT SLOUGH BRIDGE
Located at the south end of the Old Hat Slough Bridge on Marine Drive. When the new bridge is finished it should provide access for a car top boat. Not suitable for trailers. CAUTION: conditions vary with tides.

WEST PASS BRIDGE
Located under the west abutment of the bridge over West Pass 0.5 miles west of Stanwood on Highway 532. Possible place to launch a car top boat. Not suitable for trailer launching. Parking is limited. Watch tide levels and soft mud.

SNOHOMISH COUNTY
SOUTH

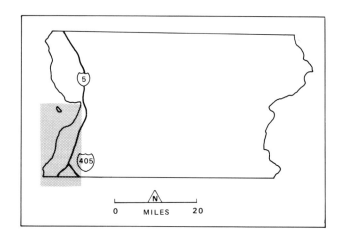
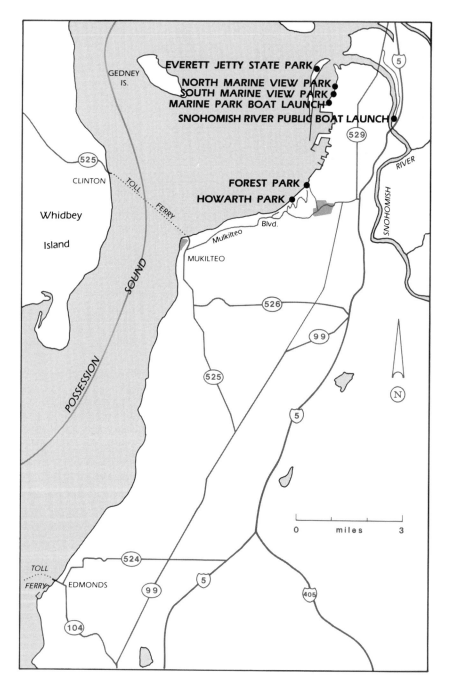

SNOHOMISH COUNTY
SOUTH

PUBLIC SHORE

	ACRES	CAMP UNITS	PICNIC UNITS	RESTROOMS	FIREPITS	SWIMMING BEACH	BOAT LAUNCH (lanes)	BOAT MOORAGE (slips/buoys)	PUBLIC PIER	DRINKING WATER	VIEWPOINT	SHORELINE LENGTH (feet)	ROCK BEACH	SAND BEACH	GRAVEL BEACH	MUD BEACH	SAND DUNES	TIDE POOLS	WETLANDS	BLUFFS
EVERETT JETTY STATE PARK	160.0											13,200	•							
FOREST PARK	111.0	45	•	•								3,000	•	•						
HOWARTH PARK	28.0	6	•		•						•	3,960	•	•				•	•	
MARINE PARK BOAT LAUNCH	21.3	2	•				13			•		710								
NORTH MARINE VIEW PARK	2.0										•	600								
SNOHOMISH RIVER PUBLIC BOAT LAUNCH	15.0						2					500	•	•					•	
SOUTH MARINE VIEW PARK	2.5										•	600								

EVERETT JETTY STATE PARK
Access by boat only. Park and launch at Marine Park Boat Launch and boat a short distance to the jetty. The park is undeveloped and seldom used. The openwater side is a gently sloping sandy beach. Overnight camping is allowed. No drinking water. A well known haulout area for Sea Lions. The area is known locally as "Jetty Island."

FOREST PARK
Located on Mukilteo Blvd. on the west side of Everett. Park is divided into two primary sections by Mukilteo Blvd. The upper park includes an indoor swimming pool, tennis courts, developed picnic area with covered shelters, children's animal farm, baseball field and various rental facilities. The lower park is virtually undeveloped. A paved road leads to the bottom and the shoreline. The road ends at a sewage treatment plant. The beach is not developed and not very attractive.

HOWARTH PARK
Turn off Mukilteo Blvd. on Madrona Ave. Park is on Olympic Blvd. Excellent views across the sound. Park is separated into upper and lower sections. restrooms are in the upper section. Beach is accessible from both upper and lower sections with parking lots in both areas. Beach is accessible via pedestrian trestle over railroad right-of-way. It is a good swimming beach.

MARINE PARK BOAT LAUNCH
Located on Marine View Drive West in Everett. Beach is a rock rip-rap bulkhead. Nice grassy lawn next to launch area, good for picnicking. Possibly the best designed saltwater boat launch in the state.

NORTH MARINE VIEW PARK
Located on Marine View Drive within the main industrial waterfront area of Everett. No usable beach. View of Port Gardiner Bay which is mostly of rafts of logs ready to go to the mill.

SNOHOMISH RIVER PUBLIC BOAT LAUNCH
Located on Smith Island Road up river from Highway 529. On the bank of the Snohomish River. Some sandy, gravelly beach area. Not a well maintained area, but serviceable. There are also several places to pull off on this bank of the river above and below the boat launch.

SOUTH MARINE VIEW PARK
Located on Marine View Drive within the main industrial waterfront area of Everett. Developed as a day use park in an urban setting. No usable beach. View of Port Gardiner Bay is mostly of log rafts waiting to go to the mill.

SNOHOMISH COUNTY
SOUTH

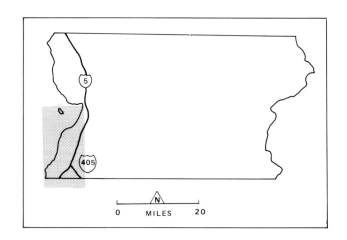

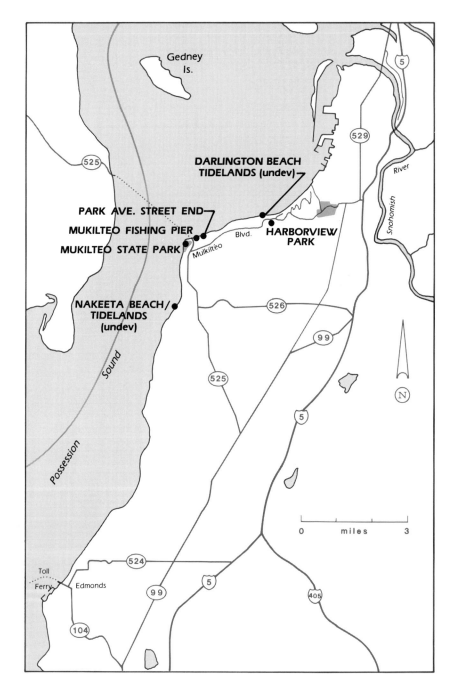

SNOHOMISH COUNTY
SOUTH

	ACRES	CAMP UNITS	PICNIC UNITS	RESTROOMS	FIREPITS	SWIMMING BEACH	BOAT LAUNCH (lanes)	BOAT MOORAGE (slips/buoys)	PUBLIC PIER	DRINKING WATER	VIEWPOINT	SHORELINE LENGTH (feet)	ROCK BEACH	SAND BEACH	GRAVEL BEACH	MUD BEACH	SAND DUNES	TIDE POOLS	WETLANDS	BLUFFS
DARLINGTON BEACH/TIDELANDS (undev)	62.3											4,600				•				
HARBORVIEW PARK	27.7	2									•	1,200								•
MUKILTEO FISHING PIER	0.1								•			6								
MUKILTEO STATE PARK	14.0	25	•	•		3					•	1,495			•					
NAKEETA BEACH/TIDELANDS (undev)	8.8											570				•				
PARK AVE. STREET END	0.5			•							•	150	•	•						

DARLINGTON BEACH/TIDELANDS (undev)
Tideflat area adjacent to Burlington Northern Railroad tracks, just north of Mukilteo. No upland public access. Includes the small delta area at the mouth of Merrill Creek.

HARBORVIEW PARK
Located on Mukilteo Blvd. near its junction with Dover Street on the southwest side of Everett. No legal access to the beach. Excellent views of Possession Sound and Port Gardner Bay. Large lawn with some fruit trees. No restroom facilities.

MUKILTEO FISHING PIER
Located adjacent to Mukilteo Ferry Terminal. Fishing pier over water provides fishing access but not beach access.

MUKILTEO STATE PARK
Located adjacent to the Mulkiteo Ferry Terminal. CAUTION: be careful when driving in the parking lot because it is laid out in reverse of logical traffic flow. Primarily a boat launching area. Beach is large cobbles; not suitable for swimming. West view of Puget Sound. Day use park, but overnight parking of boat trailers is allowed by permit (available at the boat launch).

NAKEETA BEACH/TIDELANDS (undev)
Tideflat area adjacent to Burlington Northern Railroad tracks, 1.5 miles south of Mukilteo. No developed upland public access.

PARK AVE. STREET END
Located 1 block north of the ferry terminal in Mukilteo. About 150 feet of good swimming beach. A low concrete bulkhead separates the beach from the parking area.

SNOHOMISH COUNTY
SOUTH

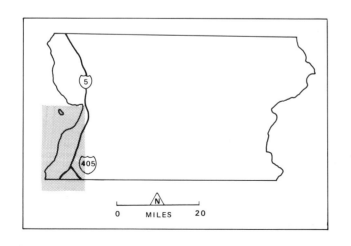
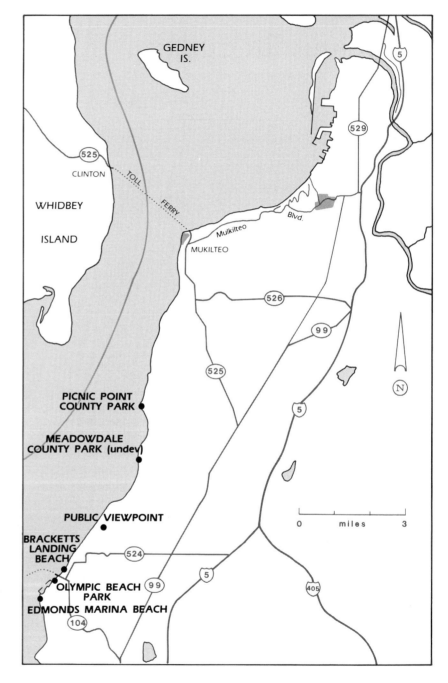

SNOHOMISH COUNTY
SOUTH

	ACRES	CAMP UNITS	PICNIC UNITS	RESTROOMS	FIREPITS	SWIMMING BEACH	BOAT LAUNCH (lanes)	BOAT MOORAGE (slips/buoys)	PUBLIC PIER	DRINKING WATER	VIEWPOINT	SHORELINE LENGTH (feet)	ROCK BEACH	SAND BEACH	GRAVEL BEACH	MUD BEACH	SAND DUNES	TIDE POOLS	WETLANDS	BLUFFS
BRACKETTS LANDING BEACH	27.0		•		•						•	1,800	•							
EDMONDS MARINA BEACH	7.0	5	•								•	978	•	•						
MEADOWDALE COUNTY PARK (undev)	100.0											840	•	•				•		
OLYMPIC BEACH PARK	4.0	4	•						•		•	300	•	•						
PICNIC POINT COUNTY PARK	15.0	6	•	•							•	1,200	•	•						
PUBLIC VIEWPOINT	NA										•	NA								

BRACKETTS LANDING BEACH
Located immediately north of the Kingston Ferry Terminal in Edmonds. Popular scuba diving area. Nice sandy beach in a protected location. No collecting of marine life. Underwater park with sunken artificial habitats. There is a scenic viewpoint on Sunset Ave. above the park and across the railroad tracks which is a separate park. Ranger is only on site during the summer low tides.

EDMONDS MARINA BEACH
Located at the south end of Admiral Way in Edmonds. Lawn with benches that can be used as picnic tables. A nice view. Scuba diving area, but no collecting of plants and animals. Excellent location for kite flying. Park is adjacent to loading pier for nearby oil refinery. Park hours are dawn to 11:00 p.m. Ranger is on site occasionally during the summer.

MEADOWDALE COUNTY PARK (undev)
Access to the park is from the north end of 75th Place West and from the south end of 72nd Ave. West, but both are blocked. The park has been closed by the Snohomish County Council because of unsafe road conditions (slide danger and road failure).

OLYMPIC BEACH PARK
Located at the foot of West Dayton Street between the Edmonds Marina and the Kingston ferry terminal. Handicapped parking and loading zone available at the end of Beach Place Street. Very popular for picnicking and sunbathing. Fishing pier and artificial reef constructed by the Washington Department of Fisheries provides excellent fishing opportunities. Pier has a covered shelter, fish cleaning areas, and is wheelchair accessible. Also within the park is a bait shop/snack bar.

PICNIC POINT COUNTY PARK
From Highway 99 take the Picnic Point Road to its end. The developed day use area has a beach suitable for swimming. A new concrete pedestrian bridge crosses the railroad tracks to the beach. View across Puget Sound to the Olympic Mountains.

PUBLIC VIEWPOINT
Located on Olympic View Drive 1.5 miles north of Edmonds at its junction with High Street. Excellent view of Puget Sound with a bench to sit on.

FLOATS

Floating docks and pilings offer a special environment for marine creatures. Plants and animals attach to floats just as they would to rocks, but the float is independent of both the shore and the bottom, thereby offering protection from many predators. Because floats are built to rise and fall with the tide, the creatures living on them are also protected from exposure to the air but at the same time manage to get an ample supply of sunlight. This protected and specialized environment enables some species to produce particularly large and ornate individuals. Marine life on and around docks and pilings is accessible to view at all times, unlike many marine species in other locations.

Calcareous sponge

Sea anemone

Jellyfish
Aequorea aequorea

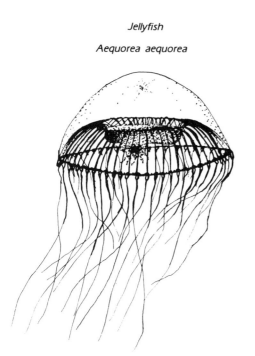

Depending on exposure to light, wave action, and freshwater supply, the flora and fauna found on a particular float will vary. Most floats will have a characteristic community established within a few months of being placed in the water.

The most obvious species one can see around a float are certain seaweeds, sponges, hydroids, sea anemones, barnacles, and mussels. In some areas, sea stars, sea urchins, and sea cucumbers move about slowly. Certain small fishes lurk in this environment but are usually difficult to see.

The variety of microscopic organisms that inhabit floats is immense. Many species have never been named or even described by professional biologists. Some of these tiny organisms can be observed with a microscope or strong magnifying glass. The microscopic life around floats is integral to the unique and complex web of life that exists in the isolated environment of floating docks.

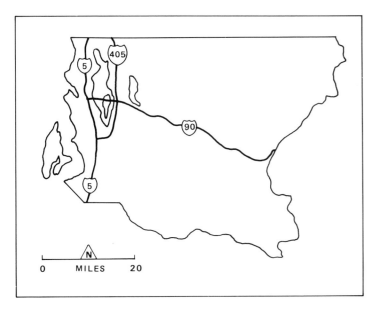

KING COUNTY

King County's ninety-one miles of shoreline are the most developed of all of the marine counties. The county is dominated by Seattle, the largest city in the state and one of the busiest seaports on the west coast. Except for the Seattle central waterfront most of King County's waterfront is residential; in places it is occupied by some of the most exclusive and expensive homes in the region. Its easy to see why with the impressive view over Puget Sound to the Olympics.

Seattle has done much in recent years to open up its central waterfront to the public and make it accessible to the adjacent business district. The Pike Place Market and historic Pioneer Square draw large crowds near the shoreline area. Between Pike Place and Pioneer Square is Waterfront Park—a combination of walkways, sitting areas, and piers that provides views of ferries, fireboats, tugboats, and large vessels in Elliott Bay. The Seattle Aquarium is just north of Waterfront Park on Pier 59. The aquarium is open seven days a week from 10 AM to 7 PM. An admission fee is charged.

The central waterfront of Seattle is made up of numerous finger piers which are historic remnants from when this area was used for shipping. Now, the shipping center has moved south to the Duawamish Waterway where modern containerized loading facilities have made the central waterfront obsolete. Several plans have been proposed for restoration of the old abandoned waterfront warehouses.

Other major features of King County's marine environment are Vashon and Maury Islands. The islands are located about halfway between Seattle and Tacoma in the middle of Puget Sound and are accessible by private boat or Washington State Ferry. Vashon Island has some nice beaches, the best of which are included in several county parks, although most of Vashon Island's shoreline is privately owned.

The entire shoreline of King County north from Seattle to Snohomish County is paralleled by a busy railroad. The railroad blocks public access and use of the shoreline except for a few developed parks where there are pedestrian overpasses.

KING COUNTY
NORTH

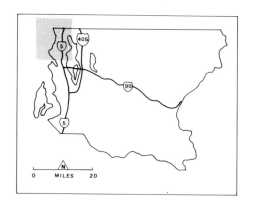

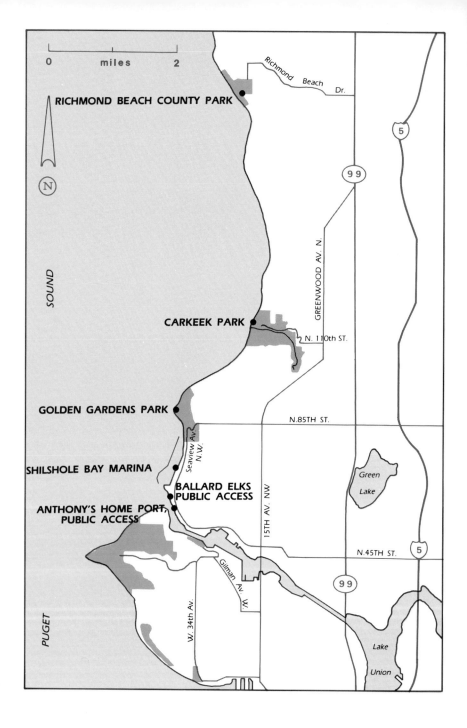

KING COUNTY
NORTH

	ACRES	CAMP UNITS	PICNIC UNITS	RESTROOMS	FIREPITS	SWIMMING BEACH	BOAT LAUNCH (lanes)	BOAT MOORAGE (slips/buoys)	PUBLIC PIER	DRINKING WATER	VIEWPOINT	SHORELINE LENGTH (feet)	ROCK BEACH	SAND BEACH	GRAVEL BEACH	MUD BEACH	SAND DUNES	TIDE POOLS	WETLANDS	BLUFFS
ANTHONY'S HOME PORT, PUBLIC ACCESS	NA										•	100								
BALLARD ELKS PUBLIC ACCESS	NA										•	100								
CARKEEK PARK	216.5	50	•	•							•	2,000	•	•						•
GOLDEN GARDENS PARK	94.6	24	•	•	•						•	3,850	•							•
RICHMOND BEACH COUNTY PARK	40.0	15	•	•	•						•	900	•							
SHILSHOLE BAY MARINA	107.7		•				4	1500	•	•	•	4,000								

ANTHONY'S HOME PORT, PUBLIC ACCESS
Located south of Shilshole Bay Marina on Seaview Drive N.W. This public access area, which is a path on the left side of the restaurant, is required by shoreline permit.

BALLARD ELKS PUBLIC ACCESS
Located on the south side of the Ballard Elks building on Seaview Ave. N.W. near Shilshole Bay. A small landscaped area with steps down the bulkhead to a sandy-gravel beach. A gravel pathway extends from the street where there is a sign marking the public access. Nice view to the west.

CARKEEK PARK
Popular waterfront and nature park in Northwest Seattle. Paths and trails vary from gently sloping to steep. Popular for picnicking and informal games and sports. Has grassy meadows, a creek, and an orchard of fruit trees. A METRO sewage treatment plant is located in the park. For group picnic reservations call 625-4671.

GOLDEN GARDENS PARK
Located at the west end of Seaview Ave. N.W., just north of Shilshole Bay Marina in the Sunset Hill District of Seattle. Nice sandy beach, a very popular area for beach combing and wind surfing. The upland area, separated from the beach by railroad tracks, is wooded and quite distant from the beach. Parking available at beach level and in the upper area. For picnic shelter reservations call 625-4671.

RICHMOND BEACH COUNTY PARK
Take Aurora Ave. north from Seattle and turn west on the N.W. Richmond Beach Road; turn left on 20th Ave. N.W. which leads directly to the park. Park sits in an old gravel borrow pit but has a very nice beach. Walkway over the railroad tracks leads to the beach. Viewpoint at the upper edge of the pit provides a view of Puget Sound.

SHILSHOLE BAY MARINA
Located on Seaview Ave. N.W.; drive through the Ballard District of Seattle and go west on N.W. Market Street. Shoreline is a rock rip-rap bulkhead. Transient moorage is available. Hoist is part of privately leased marina facilities.

KING COUNTY
NORTH

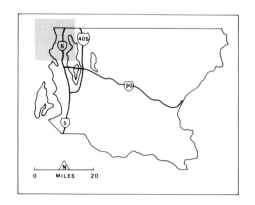

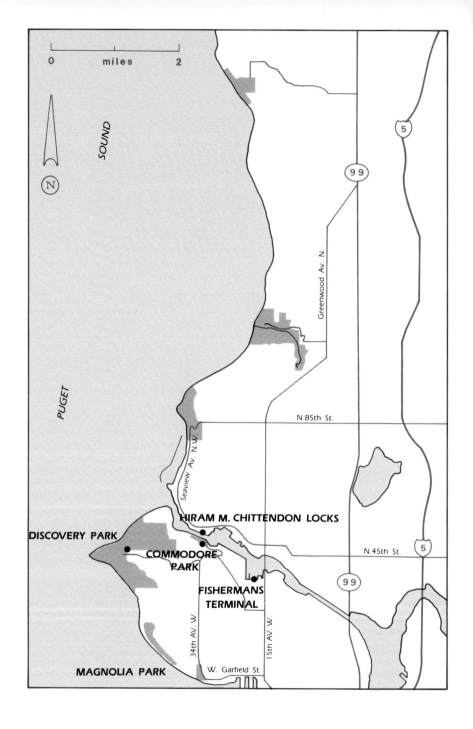

KING COUNTY
NORTH

	ACRES	CAMP UNITS	PICNIC UNITS	RESTROOMS	FIREPITS	SWIMMING BEACH	BOAT LAUNCH (lanes)	BOAT MOORAGE (slips/buoys)	PUBLIC PIER	DRINKING WATER	VIEWPOINT	SHORELINE LENGTH (feet)	ROCK BEACH	SAND BEACH	GRAVEL BEACH	MUD BEACH	SAND DUNES	TIDE POOLS	WETLANDS	BLUFFS
COMMODORE PARK	NA		5						•		•	1,200								
DISCOVERY PARK	534.0		23	•							•	12,000	•	•						•
FISHERMANS TERMINAL	76.1			•					•	•		NA								
HIRAM M. CHITTENDEN LOCKS	NA			•						•	•	2,500								
MAGNOLIA PARK	14.4		10	•	•						•	1,200			•					•

COMMODORE PARK
Located in the Magnolia/Ballard District of Seattle on West Commodore which is reached via West Emerson Place from the south end of the Ballard Bridge. The park is downstream of the Lake Washington Ship Canal fish ladder. The site has a fishing pier and picnic shelters. The shoreline is a concrete and rock rip-rap bulkhead.

DISCOVERY PARK
Located in the Magnolia District of Seattle; the south entrance is at the west end of Emerson Street; the east entrance at the end of West Government Way; the north entrance at the junction of Commodore Way and 40th West. A large natural area park which includes a fitness course, open sports field, playground equipment, trails through forested setting and to the beach, tennis courts and an arboretum of native plants. At one time it was a military base. Visitor center is near the west entrance. Most of the bluffs are wooded except along South Beach where the bluff is very shear and eroding. South Meadow was once an athletic field but is now being managed as a natural habitat.

FISHERMANS TERMINAL
Located on Salmon Bay on the Lake Washington Ship Canal in the Ballard District of Seattle. Great place to walk the floats and see the commercial fishing fleet. Interpretive signs about commercial fishing.

HIRAM M. CHITTENDEN LOCKS
Take N.W. Market Street west through the Ballard District in Seattle. The locks feature a visitor's center which is located between the north parking lot and the administration building. The center has displays on the history and operation of the canal and locks. Other attractions of the locks are fish ladders, gardens and lawns. The waterfront has rip-rap and concrete bulkheads. Also known locally as the "Ballard Locks".

MAGNOLIA PARK
Located on Magnolia Blvd. in the Magnolia District of Seattle. Beach access is from the end of 32nd Ave. West. The park is divided into two sections. Most picnic facilities are in the southern section and the best views are in the northern section. The west view overlooks Elliott Bay and the Olympics. This area has a sloping lawn with large shade trees. Bluff is wooded with no trail to the beach. Children's play area has swings only. Beach is not very attractive and not very usable at high tide. The name "Magnolia" comes from the early explorers mistaken identity of Pacific Madrona trees which are common on the bluffs in this part of Seattle.

KING COUNTY
CENTER

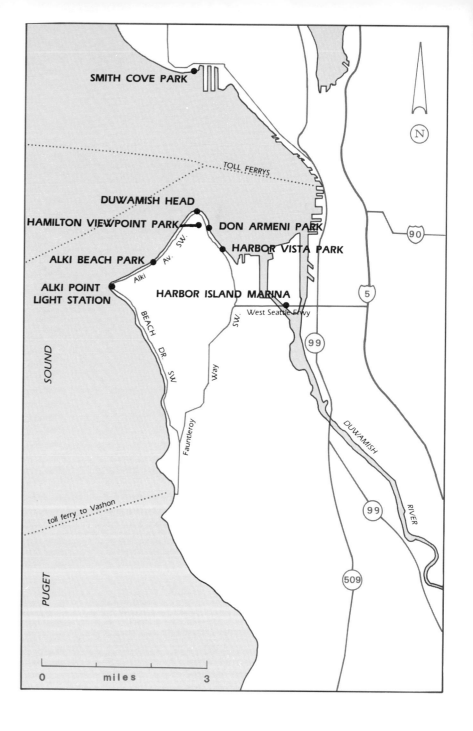
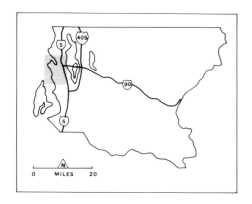

KING COUNTY
CENTER

PUBLIC SHORE

	ACRES	CAMP UNITS	PICNIC UNITS	RESTROOMS	FIREPITS	SWIMMING BEACH	BOAT LAUNCH (lanes)	BOAT MOORAGE (slips/buoys)	PUBLIC PIER	DRINKING WATER	VIEWPOINT	SHORELINE LENGTH (feet)	ROCK BEACH	SAND BEACH	GRAVEL BEACH	MUD BEACH	SAND DUNES	TIDE POOLS	WETLANDS	BLUFFS
ALKI BEACH PARK	154.1	24	•	•	•						•	13,200	•	•	•					
ALKI POINT LIGHT STATION	NA											NA								
DON ARMENI PARK	12.0	3	•				4				•	1,200								•
DUWAMISH HEAD	1.3										•	1,500								
HAMILTON VIEWPOINT PARK	NA										•	NA								
HARBOR ISLAND MARINA	NA								100	•		•	NA							
HARBOR VISTA PARK (undev)	3.9										•	1,000								
SMITH COVE PARK	2.5	7	•								•	450			•					

ALKI BEACH PARK
Located on Alki Ave. S.W. in West Seattle. Popular beach for sunbathing, walking and jogging. Large summer crowds. Play area, picnic shelters and bicycle path. The portion of the beach that is north of the lighthouse is a nice sandy beach. There is a concrete bulkhead and a paved promenade along the beach.

ALKI POINT LIGHT STATION
Located at Alki Point in West Seattle. Public visiting hours are 1:00 p.m. until 4:00 p.m. Saturday, Sunday and holidays; Wednesday - Friday is by appointment only. Closed Monday and Tuesday. No access to the beach.

DON ARMENI PARK
Located in West Seattle on Harbor Ave. S.W., overlooking Elliott Bay. Shoreline is a rock bulkhead but nicely landscaped and developed at street level. Coin operated viewing scope.

DUWAMISH HEAD
Located in West Seattle at Duwamish Head. Take West Seattle Freeway from Interstate 5; turn north on Harbor Street. Viewpoint park on a man-made fill. Some landscaping and shade trees. Beach is generally not usable; it is a rock rip-rap bulkhead.

HAMILTON VIEWPOINT PARK
Located in West Seattle near Duwamish Head on California Ave. S.W. Offers an excellent view of Elliott Bay and downtown Seattle. Coin operated telescope. No beach, viewpoint only.

HARBOR ISLAND MARINA
Located on Harbor Island immediately south of the West Seattle Freeway Bridge. Drive under the bridge to reach the marina. A public viewing pier and a float with boat fuel and groceries.

HARBOR VISTA PARK (undev)
Located in West Seattle on Harbor S.W. overlooking Elliott Bay. View of Seattle across the bay. Has parking and a bait shop. Former site of Seacrest Marina. Shoreline is a rock bulkhead with no usable beach.

SMITH COVE PARK
Located in the Magnolia District of Seattle; from 15th Ave. W. go westbound on W. Davis to 20th Ave. W.; turn left on 20th which turns into Thorndyke Ave. W.; turn left on 21st Ave. W.; follow 21st Ave. to its end. There is a small grassy area with a few newly planted trees. Shoreline is a rock bulkhead but a gravel beach is exposed at low tide. The park is named for Dr. Henry A. Smith who settled here in 1853. Park hours are dawn to dusk.

KING COUNTY
CENTER

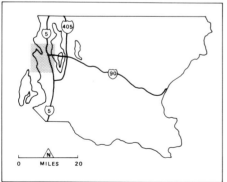

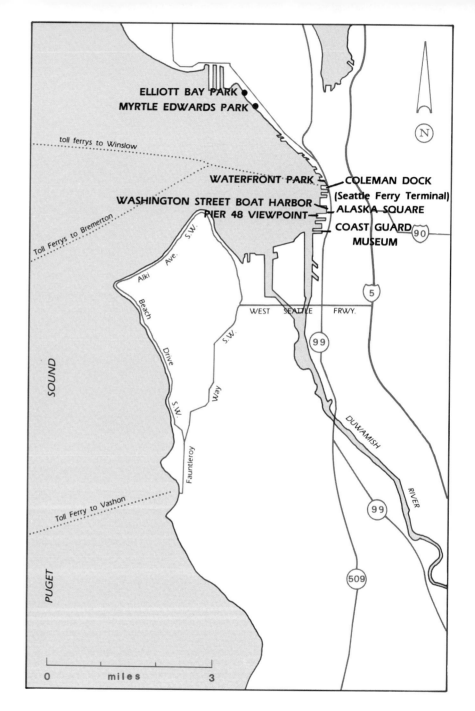

KING COUNTY
CENTER

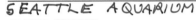
PUBLIC SHORE

	ACRES	CAMP UNITS	PICNIC UNITS	RESTROOMS	FIREPITS	SWIMMING BEACH	BOAT LAUNCH (lanes)	BOAT MOORAGE (slips/buoys)	PUBLIC PIER	DRINKING WATER	VIEWPOINT	SHORELINE LENGTH (feet)	ROCK BEACH	SAND BEACH	GRAVEL BEACH	MUD BEACH	SAND DUNES	TIDE POOLS	WETLANDS	BLUFFS
ALASKA SQUARE	0.2								•		•	225								
COAST GUARD MUSEUM	NA		•									NA								
COLEMAN DOCK (Seattle Ferry Terminal)	NA		•								•	NA								
ELLIOTT BAY PARK	10.0	1	•						•		•	4,100								
MYRTLE EDWARDS PARK	4.3		•									1,600	•							
PIER 48 VIEWPOINT	NA										•	NA								
WASHINGTON STREET BOAT HARBOR	NA							6				NA								
WATERFRONT PARK	10.4	20	•						•		•	50								

★ SEATTLE AQUARIUM

ALASKA SQUARE
Located on the central waterfront of Seattle adjacent to Pier 48. This is the small landscaped area next to the Washington Street Boat Harbor.

COAST GUARD MUSEUM
Located at Alaska Way South and South Atlantic Street in the heart of Seattle's industrial waterfront. Museum exhibits. No beach access. Public visiting hours are noon - 6:00 p.m. Monday - Saturday (open on a volunteer basis so hours may vary).

COLEMAN DOCK (Seattle Ferry Terminal)
Located on the central waterfront of Seattle. Ferries to Bremerton and Winslow. Old Coleman Dock interpretive exhibit inside the terminal.

ELLIOTT BAY PARK
Located west of Seattle's Space Needle. From Elliott Ave. turn west on Galer, cross the railroad tracks and turn left on 16th Ave. to parking at the end of the street; pathway leads to the park. Disabled access and parking can be reached by turning left on the road to Cargill Grain Terminal and taking the gravel road through chain link fences to the bait shop. Ignore the "No trespassing" sign. Waterfront is a rock bulkhead. Nice landscaped lawns extend along the waterfront strip. Fishing pier hours are 7:00 a.m. - 11:00 p.m. during the winter and 6:00 a.m. - midnight during the summer. Views to the west of the Olympic Mountains. Connects directly with Myrtle Edwards Park.

★ The south apron too!

MYRTLE EDWARDS PARK
Located in Seattle immediately north of the central waterfront area. Park in the lot at the north end of Alaskan Way. Shoreline is a rock rip-rap bulkhead. The pedestrian path and the bikeway are both 1.25 miles long. The fishing pier at Elliott Bay is located 1 mile north of Myrtle Edwards. This shoreline is also accessible for fishing. Popular summer lunch-hour site for workers from nearby Seattle office buildings. Chemical pit toilets.

PIER 48 VIEWPOINT
Located on Alaskan Way at Pier 48, the Alaskan Marine Highway Terminal. Do not park in the pier 48 parking lot. Viewing periscopes provide views to the working port area where visitors can see bulk containers being loaded onto ships. Interpretive panels offer information about history and industrial waterfront uses.

WASHINGTON STREET BOAT HARBOR
Washington Street Dock is a day-moorage facility with slips marked for 4-hour and 8-hour tie-ups. It is located at the foot of Washington Street adjacent to Alaska Square, a small landscaped area.

WATERFRONT PARK
Located on the central waterfront in Seattle near pier 57. The park has 200 feet of pier for fishing, elevated walkways and street level boardwalks adjacent to the Seattle Aquarium. This is a popular waterfront area.

KING COUNTY
CENTER

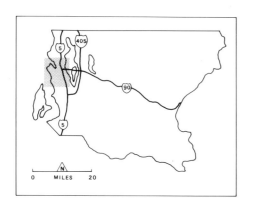
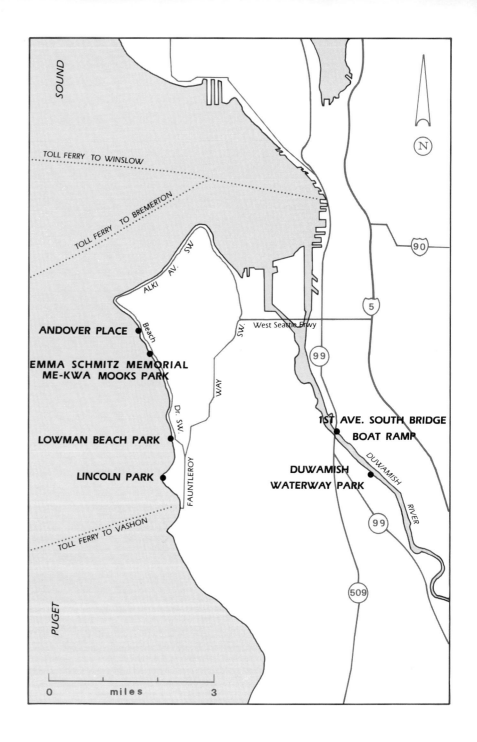

KING COUNTY
CENTER

	ACRES	CAMP UNITS	PICNIC UNITS	RESTROOMS	FIREPITS	SWIMMING BEACH	BOAT LAUNCH (lanes)	BOAT MOORAGE (slips/buoys)	PUBLIC PIER	DRINKING WATER	VIEWPOINT	SHORELINE LENGTH (feet)	ROCK BEACH	SAND BEACH	GRAVEL BEACH	MUD BEACH	SAND DUNES	TIDE POOLS	WETLANDS	BLUFFS
1ST AVE. SOUTH BRIDGE BOAT RAMP	0.4						2					30								
ANDOVER PLACE	0.4											20			•					
DUWAMISH WATERWAY PARK	1.0	1								•		200			•					
EMMA SCHMITZ MEMORIAL/ME-KWA MOOKS PARK	34.2		3							•		2,000			•					
LINCOLN PARK	130.0		282	•	•					•		5,350	•	•					•	
LOWMAN BEACH PARK	4.2			•						•		400			•					

1ST AVE. SOUTH BRIDGE BOAT RAMP
Located at the end of S. River Street under the 1st Avenue South Bridge on the Duwamish Waterway. Boat launch ramps are steep and parking under the bridge is limited.

ANDOVER PLACE
In the right-of-way of Andover Place this facility provides a trail to Puget Sound.

DUWAMISH WATERWAY PARK
Located on the Duwamish Waterway, this small neighborhood park offers an open space for waterfront viewing, picnicking, etc. It is owned by King County and managed by Seattle Parks and Recreation.

EMMA SCHMITZ MEMORIAL/ME-KWA MOOKS PARK
Located in West Seattle on Beach Drive S.W. The only parking is on the street. The beach is not very usable; there are big cobbles and seaweed. The upland area is mostly undeveloped with forested slopes. Emma Schmitz Memorial Overlook provides views to the west of the Olympic Mountains.

LINCOLN PARK
Located on Williams Point in West Seattle, immediately north of Fauntleroy Cove. Beach has a concrete bulkhead to protect against wave erosion. Most of the developed park is at the top of the bluff about 125 feet above sea level. There are large expanses of lawns with a variety of shade trees plus forested areas. View across Puget Sound of the Olympic Mountains. Features of the park are open sports fields, a developed picnic area with covered shelters, a children's play area, an outdoor fitness course, tennis courts and bicycle and walking trails. There is a swimming pool at beach level. For picnic shelter reservations call 625-4671.

LOWMAN BEACH PARK
Located north of Lincoln Park at the junction of Lincoln Parkway S.W., 48th S.W. and Beach Drive S.W. The park has a tennis court, children's swings and a large lawn with big shade trees. There is no bank at the waterfront, just a low concrete bulkhead.

KING COUNTY
SOUTH

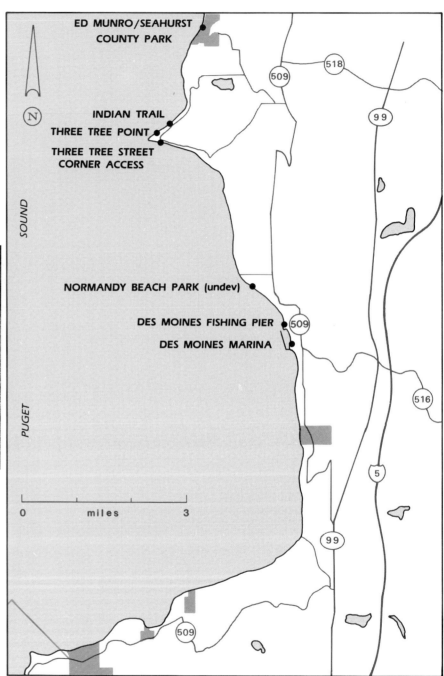

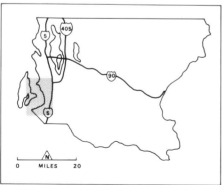

KING COUNTY
SOUTH

PUBLIC SHORE

	ACRES	CAMP UNITS	PICNIC UNITS	RESTROOMS	FIREPITS	SWIMMING BEACH	BOAT LAUNCH (lanes)	BOAT MOORAGE (slips/buoys)	PUBLIC PIER	DRINKING WATER	VIEWPOINT	SHORELINE LENGTH (feet)	ROCK BEACH	SAND BEACH	GRAVEL BEACH	MUD BEACH	SAND DUNES	TIDE POOLS	WETLANDS	BLUFFS
DES MOINES FISHING PIER	13.5		3	•					•			100								
DES MOINES MARINA	13.5			•				840				100								
ED MUNRO/SEAHURST COUNTY PARK	185.0		30	•	•	•					•	5,000	•	•						•
INDIAN TRAIL	NA										•	60			•					•
NORMANDY BEACH PARK (undev)	24.9										•	1,100	•	•						•
THREE TREE POINT	NA											40			•					
THREE TREE STREET CORNER ACCESS	NA										•	40			•					

DES MOINES FISHING PIER
Located in Des Moines on north end of Des Moines Boat Harbor. From Interstate 5 take the Kent/Des Moines exit and drive west on 516 to Des Moines; follow signs to Des Moines Marina. Utility pier provides dingy launching.

DES MOINES MARINA
Located in Des Moines. Take Des Moines exit from Interstate-5 and drive west on Highway 516 to Des Moines; follow signs to marina. Full service marina located adjacent to the fishing pier.

ED MUNRO/SEAHURST COUNTY PARK
Located in Burien in West Seattle. From the junction of Highways 518 and 509 take 1st Ave. south to 152nd Street; turn west on 152nd, then north on Ambaum and left on S.W. 144th Street; turn right on 13th Ave. S.W. and follow signs to park. View to the west of the Olympic Mountains. Beach is bulkheaded with gabions (rock-filled wire baskets) and concrete. Forested bluff. Landscaped promenade along the beach.

INDIAN TRAIL
The trail extends from the 16000 block of Maple Wild Ave. S.W. to the beach street end at S.W. 170th. At its intersection with 163rd Street there is a street end providing beach access. Trail is also apparently the right-of-way for a water main. This half mile trail, which leads between homes on either side, is said to be an old Indian trail. Path is closed at sunset.

NORMANDY BEACH PARK (undev)
Located on Marine View Drive S.W.; park near the gravel road that starts in the 208th block. The road has a locked cable and "no trespassing" signs. It's an undeveloped park but it provides access to a nice beach down a steep rough trail through the brush. Bluff is slowly eroding as is normal along this side of Puget Sound. Madronas and related plant community. View to the south of Mt. Rainier.

THREE TREE POINT
Drive from Burien on Sylvester Road S.W. Street end access is at the end of S.W. 120th which is west of Burien at Point Pully. Sign on the beach states that the tidelands are private and the right of access is prohibited by RCW (state law). There is a nice gravelly beach but there is not much room to park. There are "no trespassing" signs both ways.

THREE TREE STREET CORNER ACCESS
Located at the corner of S.W. 172nd Street and Maple Wild Ave. S.W. This street end provides access to extensive gravelly beach. Scenic view of Mt. Rainier to the south.

KING COUNTY
SOUTH

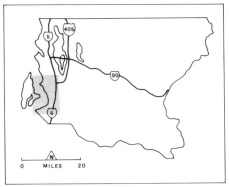

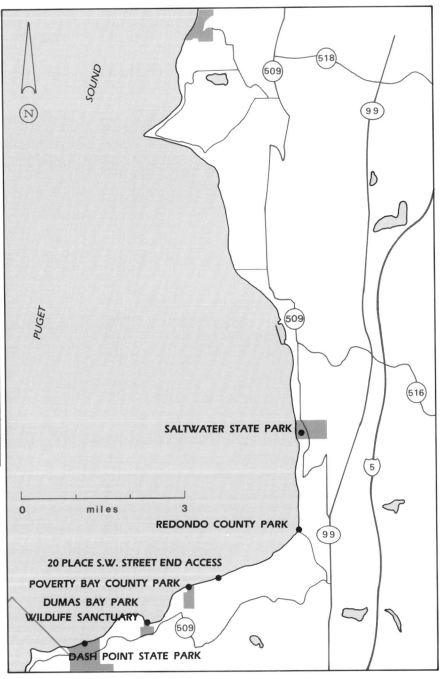

KING COUNTY
SOUTH

PUBLIC SHORE

	ACRES	CAMP UNITS	PICNIC UNITS	RESTROOMS	FIREPITS	SWIMMING BEACH	BOAT LAUNCH (lanes)	BOAT MOORAGE (slips/buoys)	PUBLIC PIER	DRINKING WATER	VIEWPOINT	SHORELINE LENGTH (feet)	ROCK BEACH	SAND BEACH	GRAVEL BEACH	MUD BEACH	SAND DUNES	TIDE POOLS	WETLANDS	BLUFFS
20 PLACE S.W. STREET END ACCESS (undev)	NA											40			•					
DASH POINT STATE PARK	395.0	163	88	•	•	•			•	•		3,000	•							•
DUMAS BAY PARK WILDLIFE SANCTUARY (undev)	23.0											450	•	•	•				•	
POVERTY BAY COUNTY PARK (undev)	37.0											1,190								
REDONDO COUNTY PARK (undev)	2.3		•				2		•			1,060			•					
SALTWATER STATE PARK	87.8	69	128	•	•			3		•	•	1,445		•						

20 PLACE S.W. STREET END ACCESS (undev)
Located at the end of 20th Place S.W.; turn west on 21st Ave. S.W. from 21st Place S.W. which turns at the signal marked S.W. 312th Street on Highway 509. Street end provides access to expanse of gravelly beach. Signs warn of private beaches.

DASH POINT STATE PARK
Located on Highway 509, 5 miles north of Tacoma. Large state in a forested setting. The nice sandy beach is popular for swimming. View across Puget Sound of Vashon and Maury islands. Park is close to urban area with shopping and restaurants.

DUMAS BAY PARK WILDLIFE SANCTUARY (undev)
Turn west on 44th Ave. S.W. from Dash Point Road. Park is undeveloped. Beach fires not permitted. Not a good beach for swimming. Nearby marsh provides habitat for wetland wildlife. Park hours are 8:00 a.m. until dusk.

POVERTY BAY COUNTY PARK (undev)
Located at the end of 302nd Place near its intersection with 30th Ave. S.W. off the Dash Point Road. Undeveloped park with no facilities.

REDONDO COUNTY PARK (undev)
Located in downtown Redondo. No upland facilities. Scuba diving area, but the beach is not usable at high tide; there is a concrete bulkhead.

SALTWATER STATE PARK
Located south of Des Moines on Highway 509. Campground is in a wooded ravine. Probably the closest public campground to Seattle. Upland hiking trails in Kent Smith Canyon and a beach for picnicking. Picnic area has covered shelters.

KING COUNTY
VASHON ISLAND

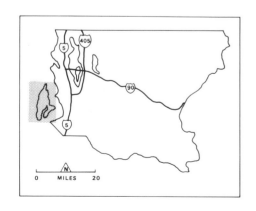
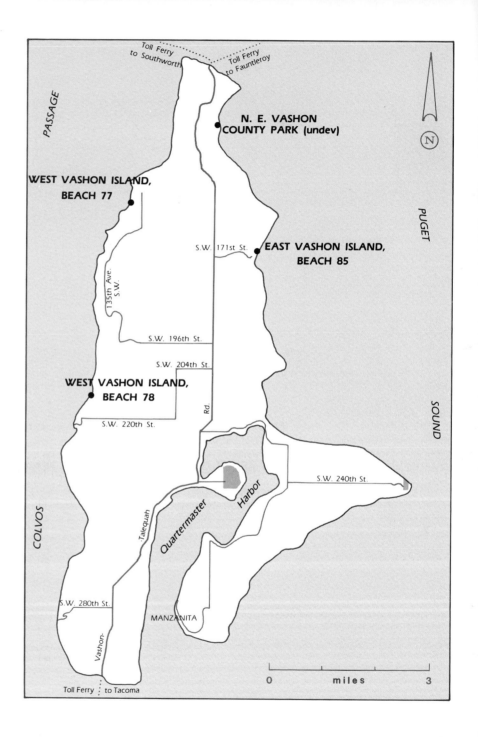

KING COUNTY
VASHON ISLAND

	ACRES	CAMP UNITS	PICNIC UNITS	RESTROOMS	FIREPITS	SWIMMING BEACH	BOAT LAUNCH (lanes)	BOAT MOORAGE (slips/buoys)	PUBLIC PIER	DRINKING WATER	VIEWPOINT	SHORELINE LENGTH (feet)	ROCK BEACH	SAND BEACH	GRAVEL BEACH	MUD BEACH	SAND DUNES	TIDE POOLS	WETLANDS	BLUFFS
EAST VASHON ISLAND, BEACH 85	NA											1,525	•		•					
N. E. VASHON COUNTY PARK (undev)	12.0											600	•		•					
WEST VASHON ISLAND, BEACH 77	NA											760	•		•					
WEST VASHON ISLAND, BEACH 78	NA											1,780	•		•					

EAST VASHON ISLAND, BEACH 85
Located south of Point Beals on the eastern side of Vashon Island. Access is by boat only. Beach is composed of cobbles on the upper area, sand and rock on the lower area. Only the tidelands are public.

N. E. VASHON COUNTY PARK (undev)
Located 0.5 miles south of the ferry terminal. An undeveloped park with no facilities and not suitable for public use at this time.

WEST VASHON ISLAND, BEACH 77
Located on the western side of Vashon Island south of Peter Point. Access is by boat only. Beach is composed of cobbles on the upper portion and sand and rock in the lower section. Only the tidelands are public.

WEST VASHON ISLAND, BEACH 78
Located on the western side of Vashon Island midway between the northern and southern tips of the island. Access is by boat only. The upper beach is composed of cobbles whereas the lower beach is sandy and rocky. Only the tidelands are public.

WASHINGTON STATE FERRIES

The Washington State Ferry System is the vital link that makes the state's marine shoreline accessible to everyone. With twenty-eight terminals scattered throughout Puget Sound, ferries offer enjoyable rides to the San Juan Islands, the Olympic Peninsula, and several Puget Sound islands. They also offer rides across several channels or passages that would take hours to drive around. The system is one of the largest in the world (the largest in the U.S.) and undoubtedly one of the most enjoyable public transit systems, with clean spacious cabins, cafeterias on board, and one of the best shows of the state's scenery available.

Ferries are easy to take. With the exception of the most crowded times of the year they are usually on schedule. Schedules are available at ferry terminals, at most travel information centers, and from the Washington Department of Transportation. Seattle area ferries are often crowded on weekdays during commuter hours. In the San Juans, peak travel times are Friday evening for those going westbound, and Sunday afternoon for travel eastbound. Avoiding these peak periods will reduce your wait and make your trip more relaxing and enjoyable.

When you arrive at the ferry terminal, signs or a traffic attendant will direct you to a specific lane for your destination. It's a good idea to double check your lane so you don't get shipped to the wrong destination. After loading you are free to wander around the decks, enjoy the view, and look for marine wildlife. An announcer will inform you when it is time to return to your car for unloading.

If you are traveling by foot or bicycle you will board and debark before the cars. You can park your bike in a marked area close to the head of the dock.

Rates for vehicles and passengers for all state ferry routes are available in a pamphlet entitled "Schedule of Tolls," or by calling Washington State Ferries. Lowest rates are for passenger "walk ons," bikes cost a few dollars more, and cars, recreational vehicles, and trailers are the most costly.

For information about Washington State Ferries, call one of these 24-hour numbers: Seattle area (206) 464-6400; statewide toll free 1-800-542-7052 or 1-800-542-0810.

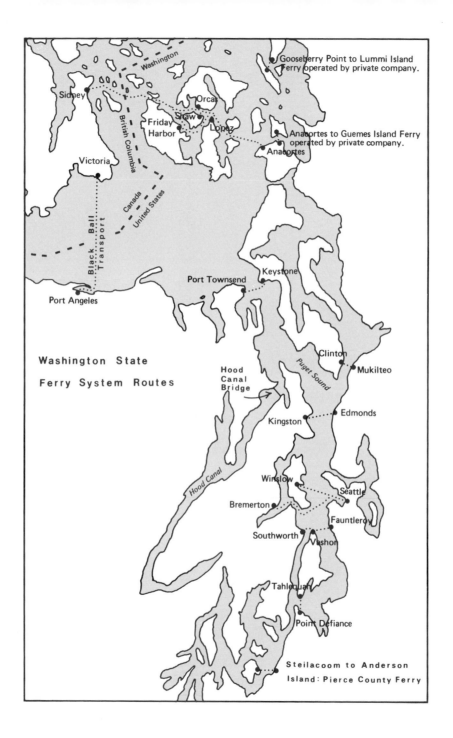

KING COUNTY
VASHON ISLAND

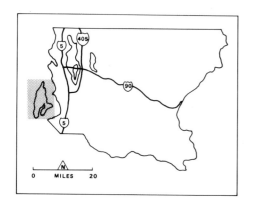

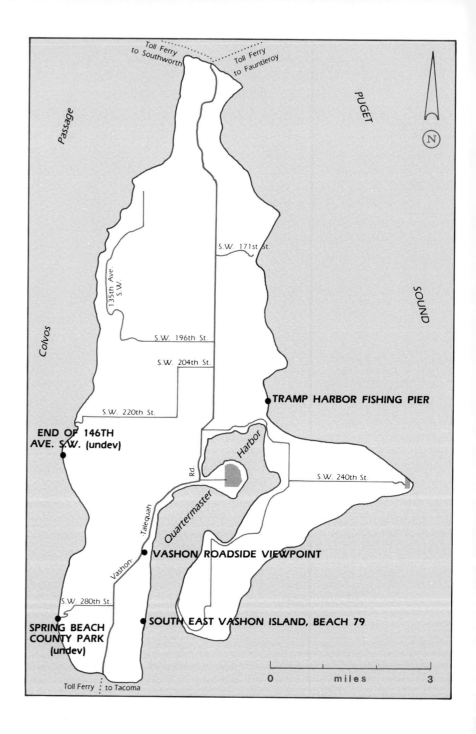

KING COUNTY
VASHON ISLAND

	ACRES	CAMP UNITS	PICNIC UNITS	RESTROOMS	FIREPITS	SWIMMING BEACH	BOAT LAUNCH (lanes)	BOAT MOORAGE (slips/buoys)	PUBLIC PIER	DRINKING WATER	VIEWPOINT	SHORELINE LENGTH (feet)	ROCK BEACH	SAND BEACH	GRAVEL BEACH	MUD BEACH	SAND DUNES	TIDE POOLS	WETLANDS	BLUFFS
END OF 146TH AVE. S.W. (undev)	NA											100	•	•						
SOUTH EAST VASHON ISLAND, BEACH 79	NA											627	•	•						
SPRING BEACH COUNTY PARK (undev)	45.8											1,300	•	•	•					
TRAMP HARBOR FISHING PIER	NA		1	•					•			50	•		•					
VASHON ROADSIDE VIEWPOINT	NA										•	NA								

END OF 146TH AVE. S.W. (undev)
Located on the west side of Vashon Island. Street end on a sandy beach near the privately operated light house.

SOUTH EAST VASHON ISLAND, BEACH 79
Located near the southern tip of Vashon Island, north of Neill Point on Quartermaster Harbor. Access is by boat only. The beach is composed of cobbles in the upper area and sand in the lower area. Only the tidelands are public.

SPRING BEACH COUNTY PARK (undev)
Located at Spring Beach on the west shore of Vashon Island. No facilities, but the site has a sandy, gravel beach.

TRAMP HARBOR FISHING PIER
Located on Vashon Island in Tramp Harbor. Park on 82nd Ave. S.W. 0.25 mile south of S.W. 209th Street. There is also a wide shoulder near 82nd Ave. S.W. and S.W. 209th Street which provides beach access. The beach is rocky.

VASHON ROADSIDE VIEWPOINT
Located on the Vashon Island Highway overlooking Quartermaster Harbor. The nice view is partly blocked by trees. A monument states: "Vashon Island, discovered 1792 by Captain George Vancouver, named for his friend Captain James Vashon. Pioneering began in 1877 when Shermans came to homestead at Quartermaster Harbor. Prices, Miners and Gilmans at paradise valley by spring 1878. All had settled on claims."

KING COUNTY
VASHON ISLAND

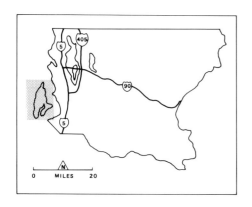
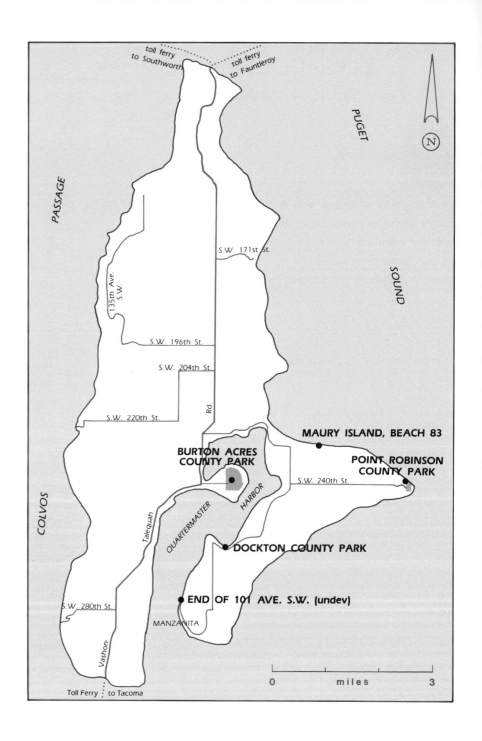

KING COUNTY
VASHON ISLAND

PUBLIC SHORE

	ACRES	CAMP UNITS	PICNIC UNITS	RESTROOMS	FIREPITS	SWIMMING BEACH	BOAT LAUNCH (lanes)	BOAT MOORAGE (slips/buoys)	PUBLIC PIER	DRINKING WATER	VIEWPOINT	SHORELINE LENGTH (feet)	ROCK BEACH	SAND BEACH	GRAVEL BEACH	MUD BEACH	SAND DUNES	TIDE POOLS	WETLANDS	BLUFFS
BURTON ACRES COUNTY PARK	57.9	2	•				1					600		•						
DOCKTON COUNTY PARK	23.0	40	•	•		•	1	10	•			1,400		•	•					
END OF 101 AVE. S.W.	NA											100		•	•					
MAURY ISLAND, BEACH 83	NA											2,000		•	•					
POINT ROBINSON COUNTY PARK	10.0		10									550		•	•					

BURTON ACRES COUNTY PARK
Take Bayview Road from the east end of S.W. 240th Street from the community of Burton on Vashon Island. Small park with a boat launch.

DOCKTON COUNTY PARK
Located on Maury Island at Stockey Road and S.W. 260th Streets near the town of Dockton. Nice park with swimming beach, lawns, forested bluff with hiking pathways; developed picnic area with a covered shelter and a children's playground. Park hours are 8:00 a.m. until dusk. Fishing pier has a mooring float. 36-hour moorage is allowed. Pier is currently closed due to disrepair.

END OF 101 AVE. S.W.
Located at the end of 101 Ave. S.W. on the south end of Maury Island on Quartermaster Harbor; county road ends near here at sign that says "Stop - Private Road." A nice sandy, gravel beach.

MAURY ISLAND, BEACH 83
Located along the northern edge of Maury Island, east of Portage. Access is by boat only. Beach is composed of cobble in the upper area and sand in the lower portions. Only the tidelands are public.

POINT ROBINSON COUNTY PARK
Located on Maury Island at the end of S.W. 240th Street; turn right on Wick Road. Nice unspoiled sandy, gravelly beach next to the U.S. Coast Guard Pt. Robinson Light Station. Small developed picnic area.

KITSAP COUNTY

Kitsap County is a peninsula bordered by Hood Canal on the west and Puget Sound on the east. It is almost entirely surrounded by water and has one of the longest saltwater shorelines, 229 miles, of any county in the state. The convoluted shore forms many inlets, channels, and bays, two large islands, Bainbridge and Blake, and many smaller islands. Although it is close to Seattle and Tacoma, Kitsap County's water borders keep it isolated and relatively undeveloped. The county can be reached by the Hood Canal Bridge from the Olympic Peninsula, the Narrows Bridge from Tacoma, and by ferry from Seattle.

Almost all of Kitsap County's beaches have stunning views of either the Olympics, on the west side, or Seattle and the Cascades on the east side. Most places along the shore also offer good opportunities to view native wildlife. Aside from an abundance of hawks and waterfowl, rare species include bald eagles and osprey. Mammal populations of bear, deer, coyote, fox, bobcat, and occasionally cougar also are present. The sharp observer can see whales, dolphins, sea lions, seals, and river otters off Kitsap County beaches.

The major population center in Kitsap County is Bremerton, which is home of the U.S. Navy's Puget Sound Naval Shipyard. Large war vessels can usually be seen from a distance here. One can board the last operating "mosquito" ferry at Bremerton and cross Sinclair Inlet to historic Port Orchard, the county seat.

North of Bremerton on the west side of the county is the Bangor Naval Base, home for Trident nuclear submarines. Aside from the base, the west side of the county is sparsely populated. One interesting town, Port Gamble northeast of the Hood Canal Bridge, has the oldest operating sawmill in America. Founded in 1853 by the Pope and Talbot Lumber Firm, its New England and Victorian homes and company store still exist as they did when the town was one of the area's major lumber ports. The old general store now houses a museum.

The town of Poulsbo on the east side of the Kitsap Peninsula has the distinct character of a small Norwegian fishing village. Two significant Indian heritage sites are also on the east side: a plaque marks the location of the Old Man House, a Suquamish longhouse estimated to have been as much as 1,000 feet long. It is also home to forty families including Chief Sealth's, for whom the city of Seattle was named. Sealth's grave is in a Suquamish cemetery nearby. The Suquamish Museum, devoted to the study of the Puget Sound Salish Indians, is just east of Poulsbo.

KITSAP COUNTY
NORTH

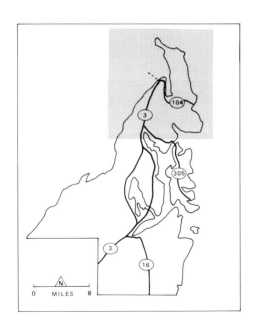
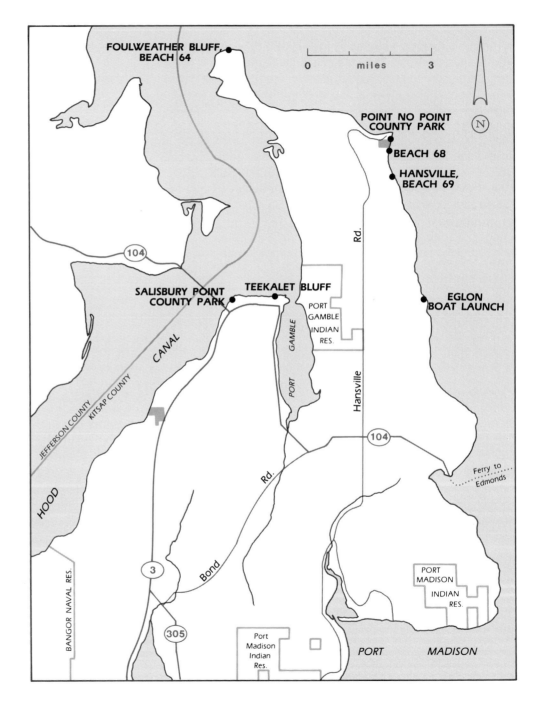

KITSAP COUNTY
NORTH

PUBLIC SHORE

	ACRES	CAMP UNITS	PICNIC UNITS	RESTROOMS	FIREPITS	SWIMMING BEACH	BOAT LAUNCH (lanes)	BOAT MOORAGE (slips/buoys)	PUBLIC PIER	DRINKING WATER	VIEWPOINT	SHORELINE LENGTH (feet)	ROCK BEACH	SAND BEACH	GRAVEL BEACH	MUD BEACH	SAND DUNES	TIDE POOLS	WETLANDS	BLUFFS
EGLON BOAT LAUNCH	NA						1					40								
FOULWEATHER BLUFF, BEACH 64	NA											3,364			•					
HANSVILLE, BEACH 69	NA											2,420		•	•					
POINT NO POINT COUNTY PARK	36.0	8	•		•						•	1,895		•	•				•	
POINT NO POINT, BEACH 68	NA											3,036		•	•					
SALISBURY POINT COUNTY PARK	5.6	10	•	•	•		3					520		•						
TEEKALET BLUFF	3.0										•	NA								•

EGLON BOAT LAUNCH
Located on the east side of Kitsap Peninsula 0.5 mile south of Eglon. No facilities other than a boat launch.

FOULWEATHER BLUFF, BEACH 64
Located at Foulweather Bluff just north of Skunk Bay. Access is by boat only. Only the tidelands are public. There is a cobble/boulder upper beach and a gravel lower beach.

HANSVILLE, BEACH 69
Located midway between Pilot Point and Point No Point on Hood Canal. Access is by boat only. There is a gravel upper beach and a sand and gravel lower beach. Only the tidelands are public.

POINT NO POINT COUNTY PARK
Take Hansville Road to Gust Halvor Road and go right; at Thors Road go left and follow gravel and sand roads into the park. Nice sandy beach; primitive. Camping is on high bluff over beaches. Dirt roads through park are impassable at some places.

POINT NO POINT, BEACH 68
Located just south of Point No Point along Hood Canal. Access is by boat only. There is a gravel upper beach and a sand and gravel lower beach. Only the tidelands are public.

SALISBURY POINT COUNTY PARK
Take Highway 104 to 0.5 mile west of Port Gamble; located off Bridge Way Road. Day use only. View of the Olympics and Hood Canal Bridge.

TEEKALET BLUFF
Take Highway 104 north to Port Gamble. Located on the north shoreline of Port Gamble, near the historical sites. Grassy open bluff overlooking the water. No facilities and no access to the water.

KITSAP COUNTY
NORTH

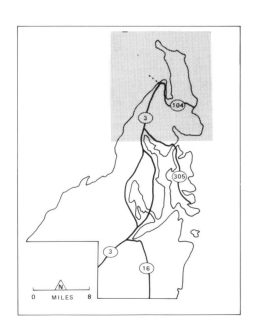
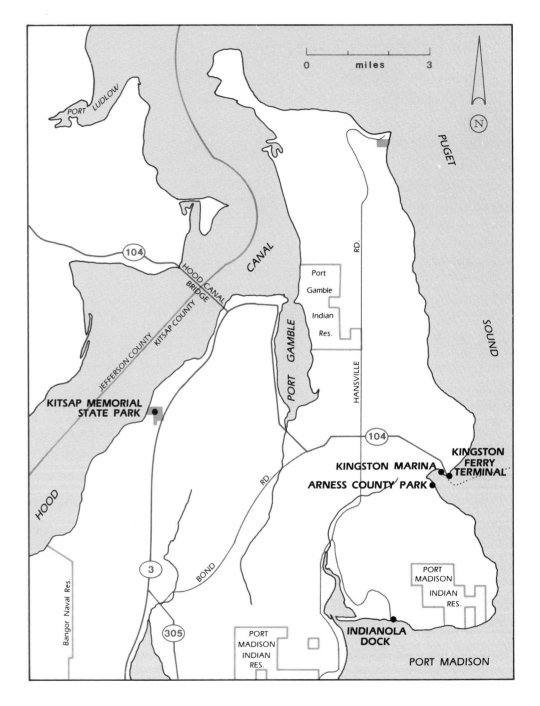

KITSAP COUNTY
NORTH

	ACRES	CAMP UNITS	PICNIC UNITS	RESTROOMS	FIREPITS	SWIMMING BEACH	BOAT LAUNCH (lanes)	BOAT MOORAGE (slips/buoys)	PUBLIC PIER	DRINKING WATER	VIEWPOINT	SHORELINE LENGTH (feet)	ROCK BEACH	SAND BEACH	GRAVEL BEACH	MUD BEACH	SAND DUNES	TIDE POOLS	WETLANDS	BLUFFS
ARNESS COUNTY PARK	2.0		5	●	●	●					●	400	●	●						
INDIANOLA DOCK	0.8								●			40								
KINGSTON FERRY TERMINAL	NA			●							●	NA	●	●						
KINGSTON MARINA	10.0		5	●				264		●		400						●		
KITSAP MEMORIAL STATE PARK	57.6	56	72	●	●	●			●	●		1,797	●		●					●

ARNESS COUNTY PARK
Located 0.5 mile outside of Kingston on South Kingston Road, across the bay from the marina and ferry dock. Roadside pulloff, picnic tables and view of Kingston.

INDIANOLA DOCK
Located in the town of Indianola. Limited parking. Pier for walking and fishing. During mid-summer a float is provided for loading and unloading of small vessels, but no mooring is allowed.

KINGSTON FERRY TERMINAL
Limited facilities at ferry terminal, trail to the beach.

KINGSTON MARINA
Located in Kingston next door to the ferry terminal. Full service marina with a boat hoist. Small grassy day use park, but no usable beach. Transient moorage available.

KITSAP MEMORIAL STATE PARK
Located 3 miles north of Poulsbo off Highway 3. Saltwater frontage on Hood Canal. Activities in the park include fishing, hiking, boating, clamming and scuba diving. The day use area also has picnic tables, barbecue braziers, two kitchen shelters, a baseball field, volleyball court, horseshoe pit and swings. A large community house can be reserved for groups. Children's playground equipment donated by the local community in 1984.

KITSAP COUNTY
CENTER

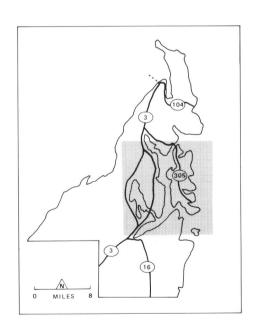

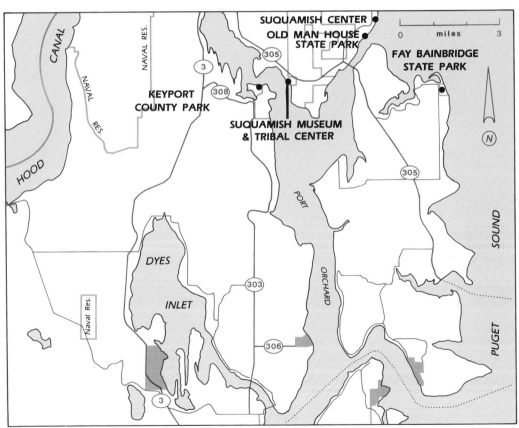

KITSAP COUNTY
CENTER

	ACRES	CAMP UNITS	PICNIC UNITS	RESTROOMS	FIREPITS	SWIMMING BEACH	BOAT LAUNCH (lanes)	BOAT MOORAGE (slips/buoys)	PUBLIC PIER	DRINKING WATER	VIEWPOINT	SHORELINE LENGTH (feet)	ROCK BEACH	SAND BEACH	GRAVEL BEACH	MUD BEACH	SAND DUNES	TIDE POOLS	WETLANDS	BLUFFS
FAY BAINBRIDGE STATE PARK	16.8	36	80	•	•	•	2			•		1,420	•	•						
KEYPORT COUNTY PARK	1.7			•			1			•		150			•				•	
OLD MAN HOUSE STATE PARK	0.7			•	•	•			•	•		210	•	•						
SUQUAMISH CENTER	4.0						1		•			1,500								
SUQUAMISH MUSEUM & TRIBAL CENTER	10.0		2	•						•		1,000	•	•					•	

FAY BAINBRIDGE STATE PARK
From Highway 305 take Phelps Road to Lafayette Ave. and follow signs. A waterfront park with open field camping. Good views to the east of large ocean vessels.

KEYPORT COUNTY PARK
Located in Keyport off Grandview Blvd. There is a boat launch, but limited beach, parking and facilities.

OLD MAN HOUSE STATE PARK
Take Suquamish Way south out of Suquamish to Angeline Ave. The park is the site of Chief Selth's home and includes historical displays. Benches and grassy lawns available for picnicking. Trails to the beach. Park along the road and walk down to the park area.

SUQUAMISH CENTER
Take Suquamish Way to the small town of Suquamish. Undeveloped access area with a pier; limited beach and limited facilities.

SUQUAMISH MUSEUM & TRIBAL CENTER
Located 45 minutes from downtown Seattle via the Winslow Ferry; take Highway 305, 7 miles to Sandy Hook Road, just across Agate Pass Bridge and follow signs. Museum is in a forested setting northeast of Keyport. A Historical museum of the Puget Sound Salish people. The museum is open daily during the summer and closed monday during the winter. Hours vary by season. The area has a nature trail and a rocky beach.

KITSAP COUNTY
CENTER

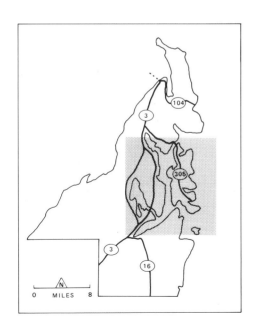
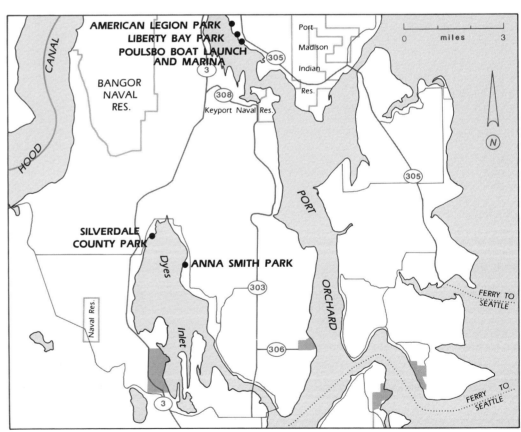

KITSAP COUNTY
CENTER

	ACRES	CAMP UNITS	PICNIC UNITS	RESTROOMS	FIREPITS	SWIMMING BEACH	BOAT LAUNCH (lanes)	BOAT MOORAGE (slips/buoys)	PUBLIC PIER	DRINKING WATER	VIEWPOINT	SHORELINE LENGTH (feet)	ROCK BEACH	SAND BEACH	GRAVEL BEACH	MUD BEACH	SAND DUNES	TIDE POOLS	WETLANDS	BLUFFS
AMERICAN LEGION PARK	3.4	3	•						•	•		1,100	•	•	•	•			•	
ANNA SMITH PARK	7.5											680		•					•	
LIBERTY BAY PARK	2.5	6	•	•	•			40	•	•		1,438	•							
POULSBO BOAT LAUNCH AND MARINA	NA		•				1		•	•		NA								
SILVERDALE COUNTY PARK	4.0	4	•		•	3				•		600	•	•	•	•				

AMERICAN LEGION PARK
Located just north of Poulsbo City Center on Front Street. Nice trails along the waterfront and a hike down to the beach.

ANNA SMITH PARK
Located 1.5 miles from Silverdale; go east on Bucklin Hill Road out of Silverdale, turn right on Tracyton Boulevard to the intersection of Fairgrounds Road where the park is located on the waterfront. No public facilities.

LIBERTY BAY PARK
Take Front Street to Poulsbo City Center; park is along the waterfront. The park has an arboretum, paved trails, and a boardwalk through trees along the water.

POULSBO BOAT LAUNCH AND MARINA
Located in downtown Poulsbo. Facilities include permanent and transient moorage, boat launch and shower facilities.

SILVERDALE COUNTY PARK
Going north or south on Highway 3, turn off at the Newberry Hill Exit to Silverdale Way, north to Byron Street in Silverdale, and follow signs. From Buklin Hill Road (from East Bremerton) turn south on Bayshore Drive; take Washington Ave. to the end. A grassy park on a quiet bay with nice picnic facilities.

KITSAP COUNTY
CENTER

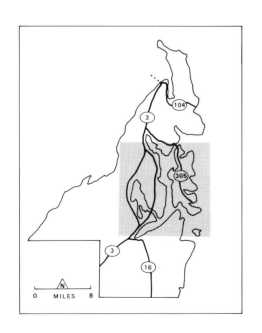

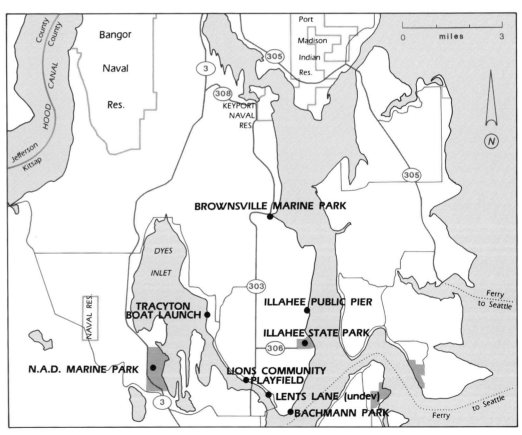

KITSAP COUNTY
CENTER

	ACRES	CAMP UNITS	PICNIC UNITS	RESTROOMS	FIREPITS	SWIMMING BEACH	BOAT LAUNCH (lanes)	BOAT MOORAGE (slips/buoys)	PUBLIC PIER	DRINKING WATER	VIEWPOINT	SHORELINE LENGTH (feet)	ROCK BEACH	SAND BEACH	GRAVEL BEACH	MUD BEACH	SAND DUNES	TIDE POOLS	WETLANDS	BLUFFS
BACHMANN PARK	0.5		2									70	•		•					
ILLAHEE PIER	4.0							6	•		•	100	•		•					
ILLAHEE STATE PARK	74.9	41	88	•	•	1		19	•	•	•	1,785	•		•					•
LENTS LANE (undev)	1.5										•	30	•							
LIONS COMMUNITY PLAYFIELD	15.0		6	•	•	3			•	•	•	1,700	•		•					
N.A.D. MARINE PARK	27.5											800	•							•
PORT OF BROWNSVILLE MARINE PARK AND MARINA	36.0		4	•	•		2	325	•	•	•	1,000			•	•				•
TRACYTON BOAT LAUNCH	NA						1					40								

BACHMANN PARK
Located at the end of Trenton Avenue in Bremerton. Small day use park.

ILLAHEE PIER
Located at the intersection of Illahee Road and Ocean View Road in Illahee. Pier only. Limited parking.

ILLAHEE STATE PARK
Located 3 miles northeast of Bremerton near Highway 306; from 306 go east on Sylvana Way to the park. Part of the park was developed under the Civilian Conservation Corps (CCC) in the 1930's and some of the structures remain. Fishing, boating, water-skiing, swimming, clamming and crabbing are all popular activities in the area. Camping in the park is on the bluff above the beach. It is a steep hike or drive down to the beach.

LENTS LANE (undev)
Located on Lebo Blvd. in East Bremerton. This is a small site near the Highway 303 bridge that has no facilities.

LIONS COMMUNITY PLAYFIELD
Located on Lebo Blvd. east of Highway 303 in Bremerton, along the Washington Narrows. The park has baseball fields, an excellent children's play area, and a running track and work-out area. Part of the Lebo Recreation Area.

N.A.D. MARINE PARK
Take Highway 3 out of Bremerton To Kitsap Way and go west; At Shorewood Drive turn right to park. Park has a jogging trail but no facilities. A gate closes off the park to vehicles, but you can walk along the road near the shore. Wooded park with a rocky shore.

PORT OF BROWNSVILLE MARINE PARK AND MARINA
Located off S.R. 303, 8 miles northeast of Bremerton. The picnic area is on the hill overlooking the marina. The site has boat rentals, live bait, a marine railway, a fishing pier, boat moorage, and a store.

TRACYTON BOAT LAUNCH
From Central Valley Road which goes through Tracyton, turn west on Tracy Ave. and follow to the end. No facilities other than the launching ramp.

KITSAP COUNTY
CENTER

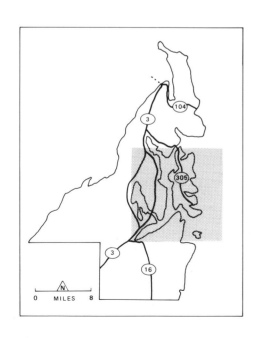
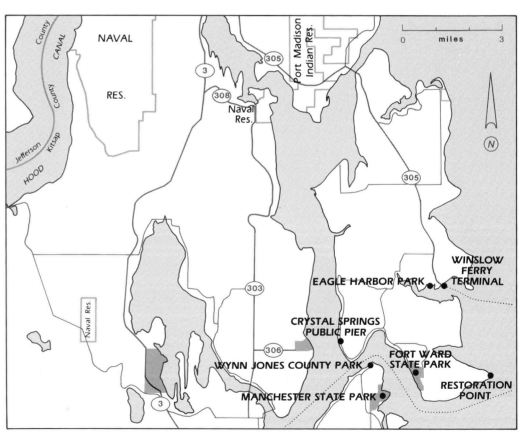

KITSAP COUNTY
CENTER

PUBLIC SHORE

	ACRES	CAMP UNITS	PICNIC UNITS	RESTROOMS	FIREPITS	SWIMMING BEACH	BOAT LAUNCH (lanes)	BOAT MOORAGE (slips/buoys)	PUBLIC PIER	DRINKING WATER	VIEWPOINT	SHORELINE LENGTH (feet)	ROCK BEACH	SAND BEACH	GRAVEL BEACH	MUD BEACH	SAND DUNES	TIDE POOLS	WETLANDS	BLUFFS
CRYSTAL SPRINGS PUBLIC FISHING PIER	NA								•			NA								
EAGLE HARBOR PARK	8.0		16	•		1				•		1,000	•	•						•
FORT WARD STATE PARK	137.1	4		•	2	2			•			4,300	•		•					
MANCHESTER STATE PARK	111.2	50	40	•	•					•		3,400	•		•			•		
RESTORATION POINT	5.0										•	1,000	•							
WINSLOW FERRY TERMINAL	NA		4	•						•		NA								
WYNN-JONES COUNTY PARK	25.0										•	250	•						•	•

CRYSTAL SPRINGS PUBLIC FISHING PIER
Located on the southwest corner of Bainbridge Island near Point White; take Point White Drive to Crystal Springs Road. A large pier only. Parking is limited. No beach.

EAGLE HARBOR PARK
On Bainbridge Island take Madison Ave. to B'June Drive; park is 0.5 mile on the right. View of ship yards and marinas in Eagle Harbor. Wooded area with a gravelly beach.

FORT WARD STATE PARK
Access to the beach is by hiking trail through the woods. The site has bird watch stations. Originally this site was a military fort.

MANCHESTER STATE PARK
Located 6 miles east of Port Orchard on Clam Bay, Rich Passage. The picnic shelter was built in 1900 to be used for torpedo storage. Many species of wildlife can be found in the park including blacktail deer, grey and red foxes, red squirrels and skunks. There are trails along the shore. Rocks along the shoreline are good fishing platforms.

RESTORATION POINT
The point of land east of Country Club, Bainbridge Island. Boat access, but watch for reefs and shallows. Foot access from the golf course.

WINSLOW FERRY TERMINAL
Located in Winslow on Bainbridge Island. Limited facilities.

WYNN-JONES COUNTY PARK
Located 3 miles east of Port Orchard on Beach Drive. Turn left on Wynn-Jones Road. Park near sign. Trail begins 200 feet up a small county road. A portion of the beach access is not yet open to the public. The viewpoint and upland trails are open.

KITSAP COUNTY
SOUTH

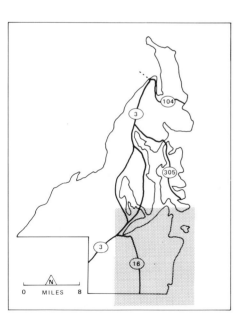
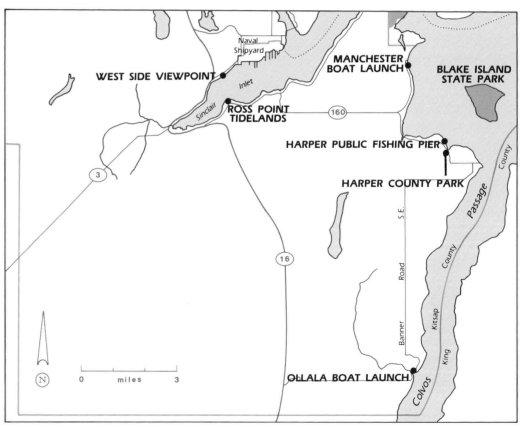

KITSAP COUNTY
SOUTH

	ACRES	CAMP UNITS	PICNIC UNITS	RESTROOMS	FIREPITS	SWIMMING BEACH	BOAT LAUNCH (lanes)	BOAT MOORAGE (slips/buoys)	PUBLIC PIER	DRINKING WATER	VIEWPOINT	SHORELINE LENGTH (feet)	ROCK BEACH	SAND BEACH	GRAVEL BEACH	MUD BEACH	SAND DUNES	TIDE POOLS	WETLANDS	BLUFFS
BLAKE ISLAND STATE PARK	475.5	36	36	•	•		43			•		17,307	•	•	•					
HARPER COUNTY PARK	40.0		•				1					500	•	•						
HARPER PUBLIC FISHING PIER	NA		•						•			30								
MANCHESTER BOAT LAUNCH	NA						1	6	•	•		40								
OLLALA BOAT LAUNCH	NA		•				1					40								
ROSS POINT TIDELANDS	3.0											650			•					
WEST SIDE VIEWPOINT	NA		2								•	NA	•							

BLAKE ISLAND STATE PARK
Located south of Bainbridge Island just a few miles from Seattle, Blake Island is accessible by boat only. The island was an ancestral camping ground for the Suquamish Indian Tribe. Four miles of beaches on the island can be reached by a trail that encircles the entire island. Activities offered on the island are boating, fishing, hiking and beachcombing.

HARPER COUNTY PARK
Located on Highway 160 south of Harper. An unkempt park with minimal development. The boat launch, for small boats, is usable only at high tide. The beach and the launch are across the road from the park.

HARPER PUBLIC FISHING PIER
Located at the intersection of Cambridge Road and Highway 160. A large newly constructed fishing pier. A view across the water to Blake Island.

MANCHESTER BOAT LAUNCH
Take Beach Drive to Manchester and turn right in town to boat launch. View of Seattle, Cascades and Blake Island. Small gravel and sand beach. Limited facilities.

OLLALA BOAT LAUNCH
Located in downtown Ollala. For small boats only. Launch into estuary and then cross under bridge.

ROSS POINT TIDELANDS
Undeveloped roadside beach area east of Port Orchard on Sinclair Inlet. No facilities.

WEST SIDE VIEWPOINT
Located at the junction of Highways 3 and 304 in South Bremerton. Small site with picnic tables and a view of Sinclair Inlet. CAUTION: beach access is across a busy road and railroad tracks.

CRABBING

Delicious Dungeness and red rock crab are both commonly harvested in Washington waters. Crabs are usually taken with crab pots, or traps, set on the bottom in relatively deep water. They may also be pulled to the surface of the water around docks and piers with bait attached to a crab line and then caught with nets. During spring and early summer, crab may be found and harvested in shallow sand and mud bays at low tide around eelgrass beds.

The **Dungeness crab**, *Cancer magister,* may reach up to eight inches across the back and has a grayish-brown shell with a purple tinge. It is found north of Seattle, particularly in the Strait of Juan de Fuca, and in Hood Canal. The Dungeness crab burrows backward into the substrate so that often only its antennae and eyes protrude. It is commonly taken commercially and sold in markets. The minimum size crab that may be harvested for food is six inches, and only males with hard shells may be kept.

The **red rock crab**, *Cancer productus,* is smaller than the Dungeness and has a heavy, brick red shell and distinctly black-tipped claws. It is more abundant intertidally than the Dungeness crab and is found throughout Puget Sound in most gravelly, muddy, and sandy bays. Because of its smaller size and heavy shell it is not harvested commercially but it is still very good to eat.

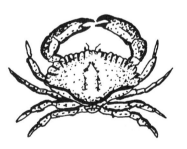

When harvesting crab, please respect the bag limit, season, and gear regulations set up by the Department of Fisheries. These regulations are published in a booklet, "Salmon, Shellfish, Marine Fish and Sport Fishing Regulations," which is available free from the Department of Fisheries, in most marinas, and in many sporting goods stores.

KITSAP COUNTY
SOUTH

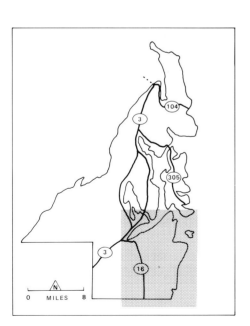
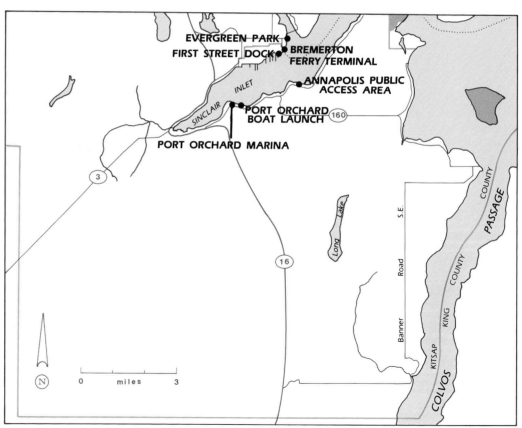

KITSAP COUNTY
SOUTH

	ACRES	CAMP UNITS	PICNIC UNITS	RESTROOMS	FIREPITS	SWIMMING BEACH	BOAT LAUNCH (lanes)	BOAT MOORAGE (slips/buoys)	PUBLIC PIER	DRINKING WATER	VIEWPOINT	SHORELINE LENGTH (feet)	ROCK BEACH	SAND BEACH	GRAVEL BEACH	MUD BEACH	SAND DUNES	TIDE POOLS	WETLANDS	BLUFFS
ANNAPOLIS PUBLIC ACCESS AREA	4.1								•			NA	•		•	•				
BREMERTON FERRY TERMINAL	NA		•									NA								
EVERGREEN PARK	6.0	10	•	•		2				•	•	300	•		•					
FIRST STREET DOCK	6.0	5					10		•			70								
PORT ORCHARD BOAT LAUNCH	1.0					2			•			100								
PORT ORCHARD MARINA	6.0		•			2	120	•	•			1,200								

ANNAPOLIS PUBLIC ACCESS AREA
Located in Annapolis next to the ferry terminal to Bremerton. Boat launch into Sinclair Inlet. New public pier. A State Department of Game Conservation License is required to use this area.

BREMERTON FERRY TERMINAL
Located in Bremerton providing ferry service to Seattle and to Port Orchard. Public facilities are available for ferry travelers.

EVERGREEN PARK
From Highway 303 in Bremerton turn east on 15th Street. Small grassy park with boat launch into Washington Narrows.

FIRST STREET DOCK
Located in downtown Bremerton at the end of 1st Street. Small paved park on water next to the ferry terminal. There are benches and picnic tables.

PORT ORCHARD BOAT LAUNCH
Located on Route 160 east of Port Orchard. Boat launch only. Small dock for loading and unloading.

PORT ORCHARD MARINA
Located in Port Orchard, next to the ferry terminal. Public restroom with shower. Free parking. Shopping area next door. There is also a boardwalk along the water.

KITSAP COUNTY
WEST

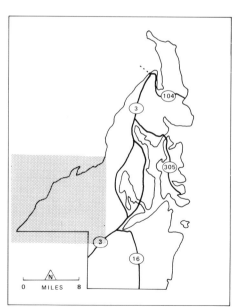
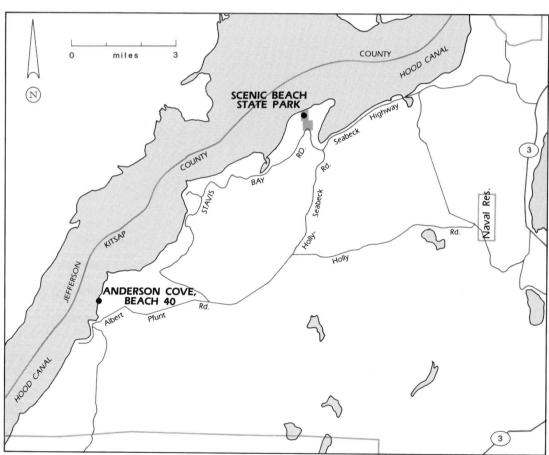

KITSAP COUNTY
WEST

	ACRES	CAMP UNITS	PICNIC UNITS	RESTROOMS	FIREPITS	SWIMMING BEACH	BOAT LAUNCH (lanes)	BOAT MOORAGE (slips/buoys)	PUBLIC PIER	DRINKING WATER	VIEWPOINT	SHORELINE LENGTH (feet)	ROCK BEACH	SAND BEACH	GRAVEL BEACH	MUD BEACH	SAND DUNES	TIDE POOLS	WETLANDS	BLUFFS
ANDERSON COVE, BEACH 40	NA											2,145	•	•	•					
SCENIC BEACH STATE PARK	71.2	50	78	•	•				•	•		1,600	•	•	•				•	

ANDERSON COVE, BEACH 40
Located on Hood Canal just south of Tekiu Point and north of Holly. Access is by boat only. The south end of the beach is sand and mud covered with eelgrass. The remainder of the beach is cobble on the upper beach and sand on the lower beach. Only the tidelands are public.

SCENIC BEACH STATE PARK
Located just south of Seabeck on Seabeck-Holly Road. Large park with excellent view of Hood Canal and the Olympic Peninsula. Grassy lawns and community center. Trails to the beach.

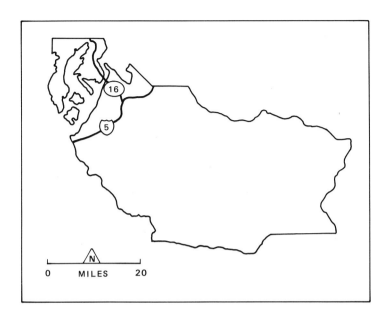

PIERCE COUNTY

Pierce County has 217 miles of saltwater shoreline. Much of the county's shore is on the Kitsap Peninsula, where numerous inlets make up a varied and scenic shoreline.

Tacoma, the second largest city in the state, is the hub of Pierce County. Mainly an industrial city which mills timber and wood pulp, Tacoma has become one of the most progressive Puget Sound communities in recent years with its waterfront improvements. Ruston Way has undergone a transformation from a rundown waterfront of obsolete terminals and dilapidated buildings to a much more friendly area with interconnected trails, parks and access sites. There is still much to be done, but Ruston Way is rapidly becoming an enjoyable place to visit.

The crown jewel of Pierce County's shoreline area is Point Defiance City Park in Tacoma. This is perhaps the most elaborately developed, yet surprisingly natural, city park on Puget Sound. Don't miss the zoo with an exhibit of polar bears and marine mammals that is second to none. There is a modest entry fee to the zoo. The main park is free. The zoo is open daily; hours fluctuate with the seasons. For more information on the Point Defiance Zoo call (206) 591-5335.

North of Tacoma, several parks provide outstanding waterfront opportunities, particularly Dash Point State Park, a major facility.

Along the stretch of beach which reaches south of Tacoma to the Fort Lewis military base there are several public access opportunities. Of special note is the pioneer community of Steilacoom, one of the state's first settlements. With stately buildings and quiet streets, Steilacoom is a favorite with many visitors. From here a ferry operates to the state prison on McNeil Island. McNeil Island is off limits to the public.

The southern boundary of Pierce County is marked by the Nisqually River Delta and National Wildlife Refuge (see Nisqually National Wildlife Refuge).

On the Kitsap Peninsula, the access opportunities of Pierce County range from simple boat launches to elaborate state parks. There is something for virtually everybody.

PIERCE COUNTY
NORTH

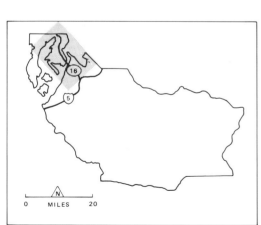

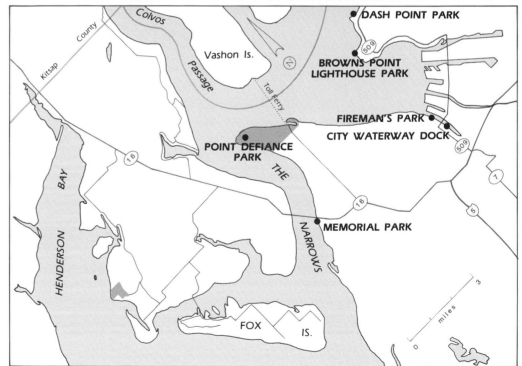

PIERCE COUNTY
NORTH

PUBLIC SHORE

	ACRES	CAMP UNITS	PICNIC UNITS	RESTROOMS	FIREPITS	SWIMMING BEACH	BOAT LAUNCH (lanes)	BOAT MOORAGE (slips/buoys)	PUBLIC PIER	DRINKING WATER	VIEWPOINT	SHORELINE LENGTH (feet)	ROCK BEACH	SAND BEACH	GRAVEL BEACH	MUD BEACH	SAND DUNES	TIDE POOLS	WETLANDS	BLUFFS
BROWNS POINT LIGHTHOUSE PARK	3.0		10	•	•	•	1			•	•		1,200	•	•	•			•		
CITY WATERWAY DOCK	NA								5	•	•		NA								
DASH POINT PARK	2.0		10	•	•					•	•	•	700	•							
FIREMAN'S PARK	1.0											•	NA								
MEMORIAL PARK	NA											•	NA								
POINT DEFIANCE PARK	700.0	300	•		•	3	10	•	•	•		12,000	•	•	•					•

BROWNS POINT LIGHTHOUSE PARK
Located at Browns Point north of Commencement Bay. Drive there via Marine View Drive. Next door is a boat launch which can be used by calling 927-7596. The park is a sloping grassy area with large shade trees. There is not a clear indication of what is public or what is private. The three acres of Browns Point offers 1200 feet of usable waterfront for fishing, boating and especially swimming and sunbathing. With all the park's amenities, the historic lighthouse and the lighthouse keeper's residence gain the most interest and attention of visitors.

CITY WATERWAY DOCK
Located in Tacoma just north of the 15th Street Bridge. Difficult to find but there is a sign which says "City Dock" on Dock Street. May be hard to find parking. The dock is a public float sandwiched between private marinas and industrial areas.

DASH POINT PARK
Located north of Tacoma. From Highway 509 take the Dash Point Road. The park is at the bottom of the hill. Small waterfront park with a narrow grass strip and beach. Most of the area is taken up with a parking lot. The fishing pier is open all year from dawn to dusk.

FIREMAN'S PARK
In Tacoma on "A" Street between S. 9th Street and S. 8th Street. The park is a lid over Schuster Parkway. Tacoma Totem Pole, the largest totem pole in the world, is located here. The totem pole, 83 feet in height, was carved by Alaskan Indians and presented to the city in 1903. View of Commencement bay and the Tacoma industrial waterfront. Historical photos on permanent display under glass. No beach, viewpoint only.

MEMORIAL PARK
Located at the east abutment of the Tacoma Narrows Bridge. Provides a scenic view of the bridge. Presently under reconstruction. No beach access.

POINT DEFIANCE PARK
Located in Tacoma at the tip of a steep-cliffed peninsula that juts into Puget Sound. Over two miles of saltwater shoreline. A swimming beach. A waterfront promenade joins with extensive trails to provide numerous hiking opportunities. Boathouse, restaurant and related facilities burned in 1984. Rebuilding plans are unknown. Extensive forest and flower gardens. Scenic overlooks provide excellent views of the water. The zoo is famous for its new Polar Bear and marine mammal exhibits (fee charged). Historic Fort Nisqually and Camp Six Museum provide interesting exhibits. One of the finest parks on Puget Sound.

PIERCE COUNTY
NORTH

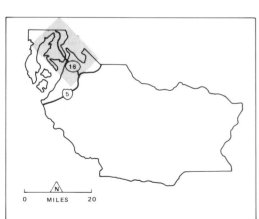

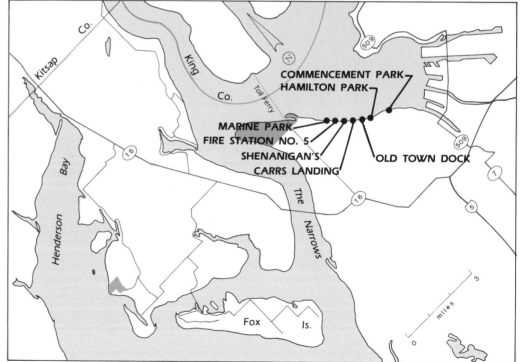

PIERCE COUNTY
NORTH

	ACRES	CAMP UNITS	PICNIC UNITS	RESTROOMS	FIREPITS	SWIMMING BEACH	BOAT LAUNCH (lanes)	BOAT MOORAGE (slips/buoys)	PUBLIC PIER	DRINKING WATER	VIEWPOINT	SHORELINE LENGTH (feet)	ROCK BEACH	SAND BEACH	GRAVEL BEACH	MUD BEACH	SAND DUNES	TIDE POOLS	WETLANDS	BLUFFS
CARRS LANDING PUBLIC ACCESS	NA										•	415	•		•					
COMMENCEMENT PARK	2.0		5	•	•				•	•	•	500	•	•						
FIRE STATION NO. 5	0.5		2								•	50	•							
HAMILTON PARK	1.0										•	300	•							
MARINE PARK	1.0		7	•							•	200	•							
OLD TOWN DOCK	NA		•						•		•	10	•							
SHENANIGAN'S PUBLIC ACCESS	NA								•		•	480	•	•						

CARRS LANDING PUBLIC ACCESS
Located on Ruston Way in Tacoma. A public access area built in conjunction with the restaurant. Pedestrian access is on the east side of the restaurant's parking lot. Access is a narrow walkway along the edge of the parking lots. Not very usable. Views of Commencement Bay. (The restaurant is presently closed. Public access areas have not been completed.)

COMMENCEMENT PARK
Located in Tacoma on Ruston Way. Linear park with benches and walkways. Rock rip-rap bulkhead. View across Commencement Bay. Settlement of Tacoma, originally "Old Town," was near here in 1865.

FIRE STATION NO. 5
Located in Tacoma on Ruston Way. View of fireboat. Next door to Katie Downs Tavern which has no public access. No beach and a rip-rap bulkhead. The park has a lawn with newly planted trees.

HAMILTON PARK
Located in Tacoma on Ruston Way west of it's intersection with McCarver. A new park which opened in 1984. Small landscaped area with some benches to sit on.

MARINE PARK
In Tacoma on Ruston Way. Nice grassy park with a bait shop and fishing pier. Scenic view of Commencement Bay and beyond. Shoreline is rock rip-rap about 10 feet high.

OLD TOWN DOCK
Located in Tacoma on Ruston Way immediately west of Commencement Park. The dock extends about 200 feet into the water. Popular fishing spot.

SHENANIGAN'S PUBLIC ACCESS
Located in Tacoma on Ruston Way. Public access is developed in conjunction with an over-water restaurant. No usable beach. Float is signed "For Shenanigan's Customers Only," but the float and pier are available for public use.

PIERCE COUNTY
NORTH

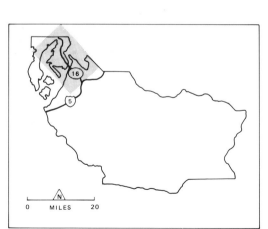
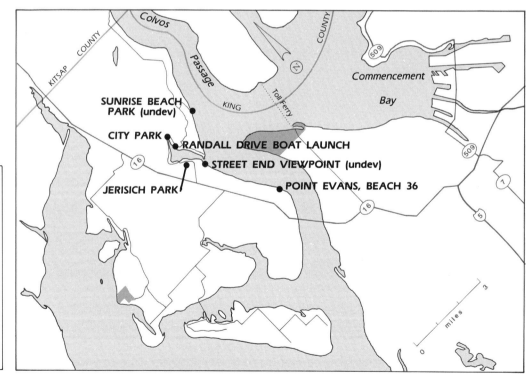

PIERCE COUNTY
NORTH

	ACRES	CAMP UNITS	PICNIC UNITS	RESTROOMS	FIREPITS	SWIMMING BEACH	BOAT LAUNCH (lanes)	BOAT MOORAGE (slips/buoys)	PUBLIC PIER	DRINKING WATER	VIEWPOINT	SHORELINE LENGTH (feet)	ROCK BEACH	SAND BEACH	GRAVEL BEACH	MUD BEACH	SAND DUNES	TIDE POOLS	WETLANDS	BLUFFS
CITY PARK	4.2	6	•	•								100						•		
JERISICH PARK	0.5							10	•			60	•							
POINT EVANS, BEACH 36	NA											2,600	•		•					
RANDALL DRIVE BOAT LAUNCH	NA						1					60								
STREET END VIEWPOINT (undev)	0.3									•	•	60								
SUNRISE BEACH PARK (undev)	30.0										•	1,200		•						•

CITY PARK
Located at the very upper end of Gig Harbor where Crescent Valley empties into the harbor. Park has a large lawn with big shade trees.

JERISICH PARK
Located in Gig Harbor at the intersection of Harborview Drive and Rosedale Street. The park has a small lawn with a bench. Beach is exposed only at low tide, otherwise there is a concrete bulkhead. Fishing pier and float to moor boats; 24 hour limit.

POINT EVANS, BEACH 36
Located just north of Point Evans which is north of the west abutment of the Tacoma Narrows Bridge. The upper beach is sand and silt and the lower beach is hardpan. Only the tidelands are public. Access is by boat only.

RANDALL DRIVE BOAT LAUNCH
Located at the east end of Gig Harbor. Launch ramp and parking only.

STREET END VIEWPOINT (undev)
Located in Gig Harbor at the east end of Harborview Drive. Viewpoint only. No apparent beach access. Parking and turn-around is a problem. View is east to Point Defiance.

SUNRISE BEACH PARK (undev)
Located on Sunrise Beach Drive N.W. which is reached via Moller Drive N.W. The beach is mostly unusable at high tide. Open grassy fields on top of the bluff. Appears to be the remnants of an old farmstead; it has not been developed much for public use. View across the sound to the smoke stacks of Tacoma. Managed as a natural area for hiking and picnicking.

PIERCE COUNTY
NORTH

Photo by Brian Walsh

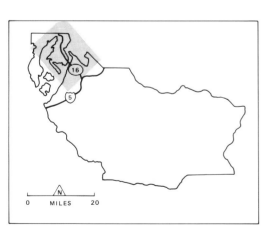

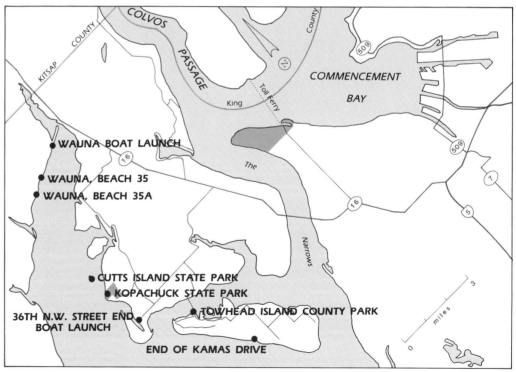

PIERCE COUNTY
NORTH

PUBLIC SHORE

	Acres	Camp Units	Picnic Units	Restrooms	Firepits	Swimming Beach	Boat Launch (lanes)	Boat Moorage (slips/buoys)	Public Pier	Drinking Water	Viewpoint	Shoreline Length (feet)	Rock Beach	Sand Beach	Gravel Beach	Mud Beach	Sand Dunes	Tide Pools	Wetlands	Bluffs
36TH N.W. STREET END BOAT LAUNCH	NA						1					60			•					
CUTTS ISLAND STATE PARK	5.5		•					9				2,740			•					
END OF KAMAS DRIVE, FOX ISLAND	NA											30			•					
KOPACHUCK STATE PARK	109.0	48	79	•	•	•				•		5,600			•					
TOWHEAD ISLAND COUNTY PARK	1.0						1			•		1,300			•					
WAUNA BOAT LAUNCH	NA					•	1			•		60								
WAUNA, BEACH 35	NA											930		•	•					
WAUNA, BEACH 35A	NA											1,504		•	•					

36TH N.W. STREET END BOAT LAUNCH
Located at the end of 36th Street N.W. on Horsehead Bay. Park on Horsehead Bay Road above. Boat launch only. Beaches on both sides are marked private.

CUTTS ISLAND STATE PARK
Located 0.5 mile offshore in Carr Inlet. Also known as Deadman's Island. Access is by boat only. Mooring buoys but no other facilities except pit toilets. No drinking water. No overnight camping. The island is believed to be an Indian burial ground. Fires are prohibited.

END OF KAMAS DRIVE, FOX ISLAND
Located at the end of Kamas Drive on Fox Island. An undeveloped street right-of-way providing limited access to the beach.

KOPACHUCK STATE PARK
Located on Henderson Bay. The park is 12 miles N.W. of Tacoma off Highway 16. Excellent beach for clamming, swimming and picnicking. Boat fishing. The name Kopachuck originated from the trade language "Chinook Jargon" of the Pacific Coastal Indians, derived from "Kopa" meaning "at" and "Chuck" meaning "water." Much of the park is a second growth forest of Douglas-fir.

TOWHEAD ISLAND COUNTY PARK
Located at the south end of the bridge to Fox Island on Hale Passage. No facilities other than the launch lane.

WAUNA BOAT LAUNCH
Located on Goodrich Drive N.W. just south of the bridge over Burley Lagoon. Boat Launch is next door to the Wauna Post Office. Parking is limited. Nearby, the south side of the west approach to the bridge provides a parking strip for access to a 0.5 mile long gravel beach.

WAUNA, BEACH 35
Located on the Longbranch Peninsula south of Burley Lagoon. Access is by boat only. Beach is composed of gravel in the upper area and sand in the lower area. Only the tidelands are public.

WAUNA, BEACH 35A
Located on the Longbranch Peninsula south of Wauna. Access is by boat only. Only the tidelands are public. Beach is composed of gravel in the upper area and sand in the lower area.

PIERCE COUNTY
NORTH

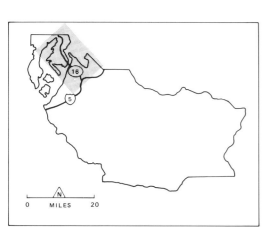

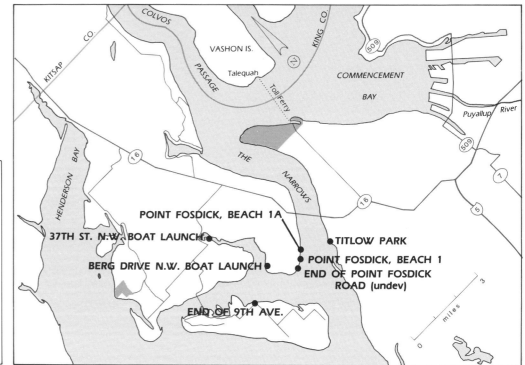

PIERCE COUNTY
NORTH

	ACRES	CAMP UNITS	PICNIC UNITS	RESTROOMS	FIREPITS	SWIMMING BEACH	BOAT LAUNCH (lanes)	BOAT MOORAGE (slips/buoys)	PUBLIC PIER	DRINKING WATER	VIEWPOINT	SHORELINE LENGTH (feet)	ROCK BEACH	SAND BEACH	GRAVEL BEACH	MUD BEACH	SAND DUNES	TIDE POOLS	WETLANDS	BLUFFS
BERG DRIVE N.W. BOAT LAUNCH	NA						1					60	•							
END OF 37TH ST. N.W. BOAT LAUNCH	NA						1					30		•						
END OF 9TH AVE., FOX ISLAND	NA								•			60		•						
END OF POINT FOSDICK ROAD (undev)	NA										•	60		•					•	
POINT FOSDICK, BEACH 1	NA											2,300			•					
POINT FOSDICK, BEACH 1A	NA											900			•					
TITLOW PARK	58.0	50	•							•	•	1,500			•					

BERG DRIVE N.W. BOAT LAUNCH
Located at the end of Berg Drive N.W. which is a short spur off of Berg Court N.W.; to get there take Pt. Fosdick Drive N.W. and turn south on 10th Street N.W. from the west end of the Tacoma Narrows Bridge. Park on the street above. Obey "no parking" signs. No facilities other than the launch.

END OF 37TH ST. N.W. BOAT LAUNCH
Located on Wollochet Bay. Take Wollochet Drive N.W. and turn on 37th Street N.W. No facilities other than the launching ramp. Limited parking.

END OF 9TH AVE., FOX ISLAND
Located on the northeast shore of Fox Island, this site is the old ferry slip. It is presently leased to a private party, but public use rights are retained.

END OF POINT FOSDICK ROAD (undev)
Located 1.0 mile south of Wollochet which can be reached by taking the airport turnoff after crossing the Tacoma Narrows Bridge (westbound). This site provides access to the beach within the road right-of-way.

POINT FOSDICK, BEACH 1
Located southwest of the west abutment of the Tacoma Narrows Bridge. Access is by boat only. There is a cobble upper beach and a gravel lower beach. Only the tidelands are public.

POINT FOSDICK, BEACH 1A
Located southwest of the west abutment of the Tacoma Narrows Bridge. Access is by boat only. The area consists of a cobble upper beach and a gravel lower beach. Only the tidelands are public.

TITLOW PARK
Located 1 mile south of the Tacoma Narrows Bridge in Tacoma. Popular scuba diving area among old pilings. Large lawns, tennis courts, basketball courts and baseball field. Fitness trail. Facilities for group events.

TIDES

In a time when tides were absent from the earth, the Raven, Klook-shood, stole the daughter of Tu-chee, the East Wind, for his wife. As a marriage present, Tu-chee promised to bare the mud-flats for twenty days so that Raven, being a shiftless fellow, could find easy food.

"Good," said the lazy raven, "but you must bare the land to the cape"; to which the East Wind replied, "No, I will make it dry for only a few feet."

The haggling went on and on until Raven finally threatened to return the daughter. Tu-chee, alarmed, compromised and agreed to make the water leave the flats twice each day. So the tides were born, and ravens and crows now go to the flat to feed.

—based on a Makah legend

Over the ages, people have looked for explanations for the mystery of the tides. Most cultures figured out that the cycles of the tide had some relationship to the sun and the moon, but the connection was not fully understood until 1687 when Sir Isaac Newton presented the law of gravitation. He discovered that the gravitational effects of the sun and the moon on the Earth's oceans cause tides. The mechanics and variations of the tides are extremely complicated, but a general explanation of the major forces clarifies the tidal phenomenon.

Gravitational force is exerted on the earth by both the sun and the moon. Even though the moon is smaller than the sun, it exerts the greater force because it is closer to the earth. As the moon orbits the earth, its gravitational pull creates a bulge in the ocean on the side of the earth closest to the moon. When this bulge hits land it makes a high tide.

At the same time, the centrifugal force created by the combined effect of the moon's gravitational pull and the earth's rotation "throws" a bulge out on the side of the earth opposite the moon. This also creates a high tide. The result is two equal but opposite forces acting on the earth creating two equal but opposite bulges in the ocean. Between the two bulges, there is a flattening effect which results in low tides. In the course of one twenty-four hour rotation, a fixed point on the earth will encounter two alternating high and low tides.

Anyone familiar with tidal cycles knows that even though the bulges may be equal, the two high tides encountered in a day at a given location are not of equal height. Every day there is a high-high tide, a low-high tide, a low-low tide and a high-low tide. This phenomenon occurs because the earth's axis of rotation is 23.5 degrees off vertical.

A fixed point rotating on the earth which travels through the thickest part of the bulge will experience high-high tide. Twelve hours later, the point will pass through a less thick part of the second bulge and experience low-high tide. The same situation occurs with low tides. In the tidal cycle of twenty-four hours fifty minutes, tides are encountered in the following order: high-high, low-low, low-high, and high-low.

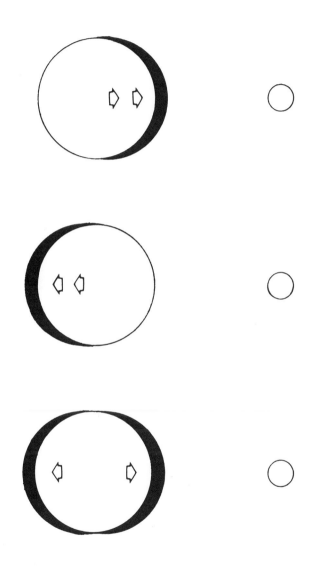

The tide cycle is slightly longer than twenty-four hours, because the moon orbits the earth every twenty-eight days. The 1,440 minutes in a twenty-four-hour day, divided by twenty-eight days, is fifty-one minutes. Fifty minutes is the amount added to the twenty-four-hour day which explains the tide interval of twenty-four hours, fifty minutes.

While the moon and the earth are producing these tides, the sun's gravity is at work producing additional, weaker tides. The sun's tides are not felt significantly except when the forces of the moon and the sun combine and create larger than normal tides. The greater than normal tides are called "spring" tides and occur when the sun and moon are aligned, during a new or full moon period. Spring tides are about 20 percent stronger than the average tide.

The opposite effect occurs when the force of the sun and moon's gravitational pull are at right angles to each other, thereby canceling each other's effect on the oceans. These tides are known as "neap" tides and occur at first and third quarter moons. Neap tides are 20 percent weaker than the average occurring tide. The major effect of the sun, therefore,

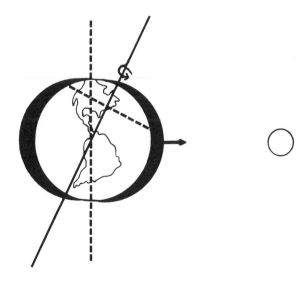

217

is to either reduce (neap) or amplify (spring) the effects of the moon on the tides twice a month.

The effect of the tidal forces produces a tidal bulge of one to two feet on the open ocean. Local geological features modify the tides, though, especially in places like lower Puget Sound, which is separated from the open sea by land masses. As a result, there is a very large difference between the high tides and low tides in Washington's inland waters.

In the Strait of Juan de Fuca a 7.2-foot tide at Cape Flattery will reach Port Townsend about three hours and forty minutes later and have a magnitude of about 7.9 feet. The time lag that occurs when tides are shaped by landforms is known as the "age of the tide." In south Puget Sound the tide is up to one hour later than at Port Townsend. The mean rise increases from 7.5 feet at Port Townsend to 13.5 feet at Olympia. In Olympia an extreme high tide of 18 feet has been recorded.

In Puget Sound the tide rushes through narrow channels and around islands, creating rapids and eddies comparable to those on a whitewater river. As any boater knows, it is important to be aware of what the tides are doing while on the water, particularly if traveling in a small boat. Some waters in the sound are not even navigable at low tide. Refer to navigation charts for more extensive tidal information in the area that you are traveling.

It is also important for visitors to the beach to be aware of tides. One could get stranded on a spit or in a cove if not aware of a rising tide. In addition, people who are interested in intertidal ecology will want to watch for the extreme low tides when the most intertidal life is exposed. Tide schedule books list high and low tides at several locations around the state. They are available at most marine stores, sporting good stores, and many general stores. In addition, many newspapers publish daily tide schedules.

DAILY TIDAL CYCLE

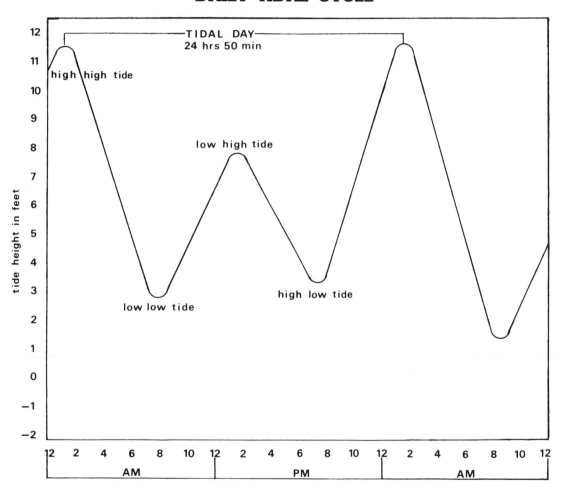

219

PIERCE COUNTY
SOUTH

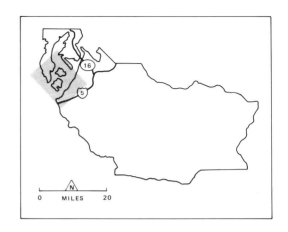
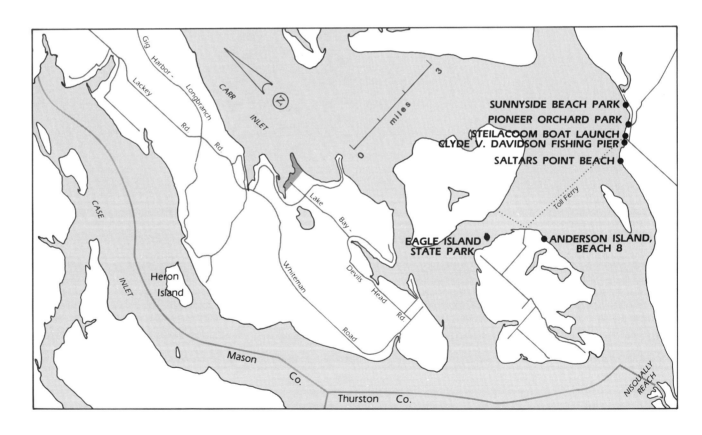

PIERCE COUNTY
SOUTH

	ACRES	CAMP UNITS	PICNIC UNITS	RESTROOMS	FIREPITS	SWIMMING BEACH	BOAT LAUNCH (lanes)	BOAT MOORAGE (slips/buoys)	PUBLIC PIER	DRINKING WATER	VIEWPOINT	SHORELINE LENGTH (feet)	ROCK BEACH	SAND BEACH	GRAVEL BEACH	MUD BEACH	SAND DUNES	TIDE POOLS	WETLANDS	BLUFFS
ANDERSON ISLAND, BEACH 8	NA											2,656		•	•					
CLYDE V. DAVIDSON FISHING PIER	NA							4	•			500								
EAGLE ISLAND STATE PARK	10.0							3				2,600	•		•					
PIONEER ORCHARD PARK	1.5										•	1,100								•
SALTARS POINT BEACH	1.1		4	•	•	•					•	500	•		•			•		
STEILACOOM BOAT LAUNCH	NA						2	4	•			50		•						
SUNNYSIDE BEACH PARK	3.0		14	•	•					•		1,400		•						

ANDERSON ISLAND, BEACH 8
Located on the eastern side of Anderson Island just south of Yoman Point. Access is by boat only. The upper beach is composed of gravel. At low tide, the lower area has sand flats. Only the tidelands are public.

CLYDE V. DAVIDSON FISHING PIER
Located off the ferry dock landing at Steilacoom. A fishing pier and an underwater artificial reef. Good bottom fishing. There is a restaurant on the dock.

EAGLE ISLAND STATE PARK
Located in Balch Passage between McNeil and Anderson islands. Boat access only. This park is intended to preserve the natural features of Eagle island so only primitive outdoor recreation is permitted.

PIONEER ORCHARD PARK
Located in Steilacoom on Commercial Street at its intersection with Wilkes Street. No beach, viewpoint only. Nice view of the sound.

SALTARS POINT BEACH
Located in Steilacoom at the corner of Champion and First Streets. Parking is limited. Footbridge over railroad tracks. Nice clean gravel beach but park facilities are in poor condition. Park closes at 10:00 PM.

STEILACOOM BOAT LAUNCH
Located under the railroad tracks at the foot of Union Avenue next to the ferry terminal. There is a charge for parking trailers and cars in addition to a launch fee.

SUNNYSIDE BEACH PARK
Located north of Steilacoom on Steilacoom Drive. Parking is removed from the site. Visitors must walk in and there is a parking charge for non-Steilacoom residents. The park has shade trees and extensive lawn. There is a sewage treatment plant next door.

PIERCE COUNTY
SOUTH

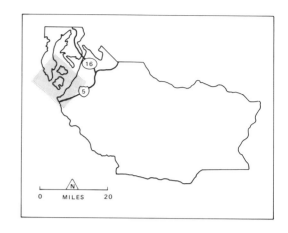
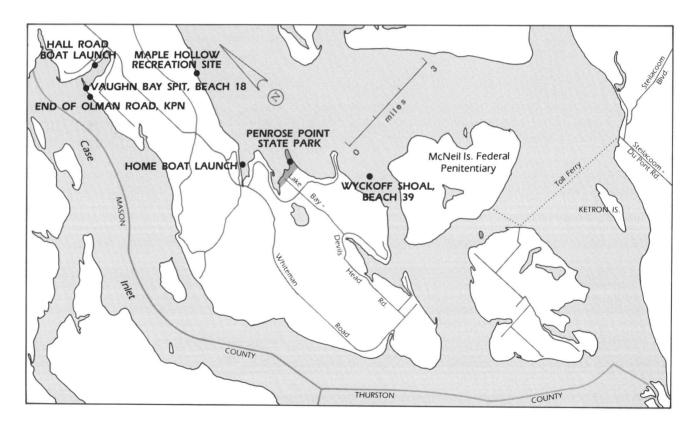

PIERCE COUNTY
SOUTH

	ACRES	CAMP UNITS	PICNIC UNITS	RESTROOMS	FIREPITS	SWIMMING BEACH	BOAT LAUNCH (lanes)	BOAT MOORAGE (slips/buoys)	PUBLIC PIER	DRINKING WATER	VIEWPOINT	SHORELINE LENGTH (feet)	ROCK BEACH	SAND BEACH	GRAVEL BEACH	MUD BEACH	SAND DUNES	TIDE POOLS	WETLANDS	BLUFFS
END OF OLMAN ROAD, KPN	NA											60			•				•	
HALL ROAD STREET END BOAT LAUNCH	NA						1					60			•					
HOME BOAT LAUNCH	NA						1					30				•				
MAPLE HOLLOW RECREATION SITE	7.0		5	•	•					•		1,420			•					
PENROSE POINT STATE PARK	145.4	95	119	•	•			16	•	•		10,076		•	•					•
VAUGHN BAY SPIT, BEACH 18	NA											1,912		•	•					
WYCKOFF SHOAL, BEACH 39	NA											NA		•	•					

END OF OLMAN ROAD, KPN
Located at the end of Olman Road near Vaughn Bay. This site is an undeveloped road right-of-way which provides access to the beach.

HALL ROAD STREET END BOAT LAUNCH
Located in Vaughn on Vaughn Bay at the end of Hall Road. No facilities other than the boat ramp.

HOME BOAT LAUNCH
Located in Home, 0.5 mile down "A" street which turns off of the Key Peninsula Highway. Provides access to Van Geldern Cove which is mostly an exposed mud flat at low tide.

MAPLE HOLLOW RECREATION SITE
Located 2.0 miles north of Home on the Key Peninsula. Turn east on Van Beek Road from the Gig Harbor/Longbranch road and drive 0.3 mile to park entrance. Day use only. Trail through deep maple forest leads down hill to the beach. A delightful place that receives minimal recreational use. Clam digging.

PENROSE POINT STATE PARK
Located on Key Peninsula 3 miles north of Longbranch on Highway 302. On Carr Inlet, much of the beach is unusable because of pile and plank bulkhead. Moorage float with overnight fee. Interpretive nature trail. Bluff is sandy loam and eroding. Hiking trails through second growth Douglas-fir forest. The Key Peninsula and surrounding Puget Sound are rich in Indian and logging history. An Indian petroglyph is in the park.

VAUGHN BAY SPIT, BEACH 18
Located at Vaughn Bay on the west side of Vaughn Bay Spit. Access is by boat only. The beach is composed of gravel with sand on top of the spit. Only the tidelands are public.

WYCKOFF SHOAL, BEACH 39
Located near Penrose Point State Park on the Longbranch Peninsula. Access is by boat only. Beach consists of coarse gravel and sand with hard clay outcrop. Only the tidelands are public.

PIERCE COUNTY
SOUTH

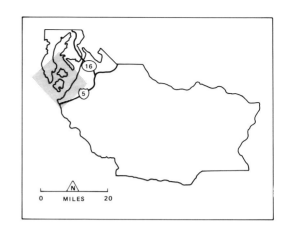

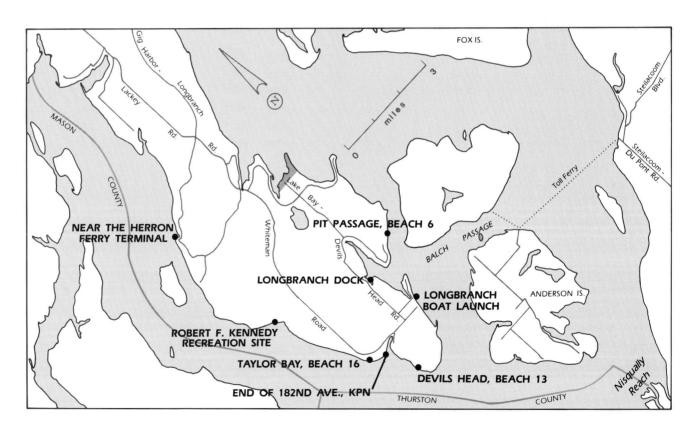

PIERCE COUNTY
SOUTH

PUBLIC SHORE

	ACRES	CAMP UNITS	PICNIC UNITS	RESTROOMS	FIREPITS	SWIMMING BEACH	BOAT LAUNCH (lanes)	BOAT MOORAGE (slips/buoys)	PUBLIC PIER	DRINKING WATER	VIEWPOINT	SHORELINE LENGTH (feet)	ROCK BEACH	SAND BEACH	GRAVEL BEACH	MUD BEACH	SAND DUNES	TIDE POOLS	WETLANDS	BLUFFS
DEVILS HEAD, BEACH 13	NA											1,344		•	•					
END OF 182ND AVE., KPN	NA										•	60			•					
LONGBRANCH BOAT LAUNCH	NA						1					60								
LONGBRANCH DOCK	NA							•	•			60								
NEAR THE HERRON FERRY TERMINAL	0.3											160			•					
PIT PASSAGE, BEACH 6	NA											1,935		•	•					
ROBERT F. KENNEDY RECREATION SITE	22.0	8	15	•	•		1	10		•		1,000			•				•	
TAYLOR BAY, BEACH 16	NA											2,500		•	•					

DEVILS HEAD, BEACH 13
Located at the southern tip of the Longbranch Peninsula. Access is by boat only. Beach is composed of gravel in the upper area and sand in the lower area. Only the tidelands are public.

END OF 182ND AVE., KPN
A public street end which provides a viewpoint overlooking Taylor Bay.

LONGBRANCH BOAT LAUNCH
Located near Longbranch at the west end of 72nd St. KPN. Boat launch facilities only.

LONGBRANCH DOCK
The old ferry slip facilities at Longbranch. Leased and managed by a private club. Moorage facilities and pier for public use.

NEAR THE HERRON FERRY TERMINAL
Located adjacent to the ferry terminal to Herron Island. The beaches west of the ferry terminal are used by the public. Parking is limited.

PIT PASSAGE, BEACH 6
Located north of Mahnckes Point on the Longbranch Peninsula. Access is by boat only. There is a gravel upper beach and a sand lower beach. Only the tidelands are public.

ROBERT F. KENNEDY RECREATION SITE
Located 5 miles south of Home on Whitman Cove. Nice quiet spot nestled in a grove of evergreen trees.

TAYLOR BAY, BEACH 16
Located on the western edge of the Longbranch Peninsula just north of Taylor Bay. Access is by boat only. Beach is composed of gravel on the upper area and sand on the lower area. Only the tidelands are public.

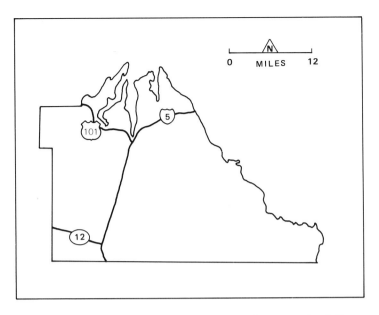

THURSTON COUNTY

Thurston County, situated at the southern end of Puget Sound is largely a region of prairies and rolling lowlands broken by minor hills and peaks. Its ninety miles of saltwater shoreline wind quietly through four fjord-like inlets—Henderson, Bud, Eld, and Totten inlets. The beaches are mostly gravel or mud, and the uplands are often high bank bluffs. The head of the inlets blend into large mudflat estuaries.

The tidal mudflats at the head of the inlets are fed by freshwater streams making them ideal environments for oyster growing. Originally all four inlets supported substantial oyster productions but currently only Totten Inlet remains unpolluted. Budd Inlet has been closed, and parts of Eld and Henderson Inlets are conditionally closed for commercial shellfish harvesting. Nevertheless, this area remains popular for recreational clam digging and shellfish gathering.

Thurston County's shoreline is almost entirely residential except for the town of Olympia. Originally named Smithfield in 1884, Olympia was one of the first pioneer settlements north of the Columbia River. In 1853 it was chosen the seat of the Washington territorial government. The state legislature now convenes in the architecturally "classic" capitol building that has one of only two solid stone domes in the United States. Tours of the State Capitol Campus are available through the Legislative Tour office. There is also a State Capitol Museum in the historic Lord Mansion which is open to the public Tuesday through Sunday.

Surrounding Olympia are several marinas with ample boating facilities and a port for large ships where logs are loaded for export across the Pacific. Olympia never became the large port it once dreamed of being, but it definitely has maintained a marine flavor.

Outside Olympia are several large county parks, a state park, and the Nisqually National Wildlife Refuge. The 2,400-acre refuge has hiking trails which give visitors the opportunity to view a wide variety of wildlife. (See Nisqually National Wildlife Refuge).

The entire south Puget Sound area is home for thousands of migratory birds in the winter. It is also heavily populated with harbor seals, and an occasional sea lion seen lounging around on a dock or pier is not uncommon.

THURSTON COUNTY

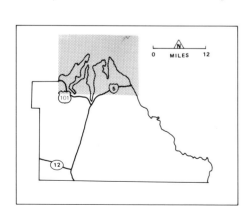
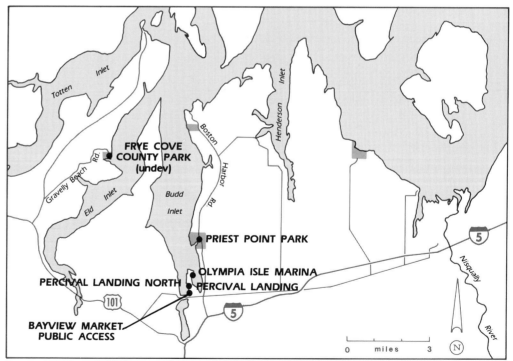

THURSTON COUNTY

PUBLIC SHORE

	ACRES	CAMP UNITS	PICNIC UNITS	RESTROOMS	FIREPITS	SWIMMING BEACH	BOAT LAUNCH (lanes)	BOAT MOORAGE (slips/buoys)	PUBLIC PIER	DRINKING WATER	VIEWPOINT	SHORELINE LENGTH (feet)	ROCK BEACH	SAND BEACH	GRAVEL BEACH	MUD BEACH	SAND DUNES	TIDE POOLS	WETLANDS	BLUFFS
BAYVIEW MARKET PUBLIC ACCESS	NA										•	100								
FRYE COVE COUNTY PARK (undev)	90.0											2,200	•	•	•			•	•	
OLYMPIA ISLE MARINA	22.0		•				2		•	•		5,000								
PERCIVAL LANDING	0.3		5	•				50	•	•		1,300								
PERCIVAL LANDING NORTH	NA										•	1,000								
PRIEST POINT PARK	253.0		125	•	•					•		3,000	•	•	•			•	•	

BAYVIEW MARKET PUBLIC ACCESS
Located in downtown Olympia just east of the 4th Ave. bridge. A small public access area located just behind the Bayview Market. View of the Port of Olympia and the Olympic Mountains.

FRYE COVE COUNTY PARK (undev)
Located off the Steamboat Island road which exits Highway 101, 5 miles northwest of Olympia. Take Gravely Beach Drive and Young Road to reach the park. Path through the woods leads to the beach. Fry cove is a small quiet inlet. Parking on the county road shoulder at the park entrance. The park is undeveloped, but it has 0.5 mile of good gravel beach on Eld Inlet.

OLYMPIA ISLE MARINA
Located in Olympia off the north end of North Washington Street. Operated by Almar Ltd. under a lease from the Port of Olympia. Transient moorage and a boat launch. Planned expansion of moorage and commercial development.

PERCIVAL LANDING
Moorage float and boardwalk on downtown Olympia waterfront. Overnight moorage, up to 72 hour stay. Porta-potty dump station for boats. Site of wooden boat fair in May.

PERCIVAL LANDING NORTH
Extends the Percival Landing boardwalk northward to the Port of Olympia. Includes a viewing tower for watching port activities (loading logs) and looking south to the state capitol building. This is a new project completed in 1985.

PRIEST POINT PARK
Located in Olympia on East Bay Drive, 2 miles north of downtown Olympia. Shady park in a Douglas-fir forest. Picnicking, playground equipment. Reservations can be made for group use of picnic shelters. Nature trail. CAUTION: beach is soft mud at low tide. View of State Capitol Building from the beach.

THURSTON COUNTY

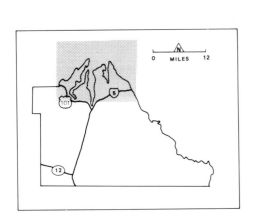
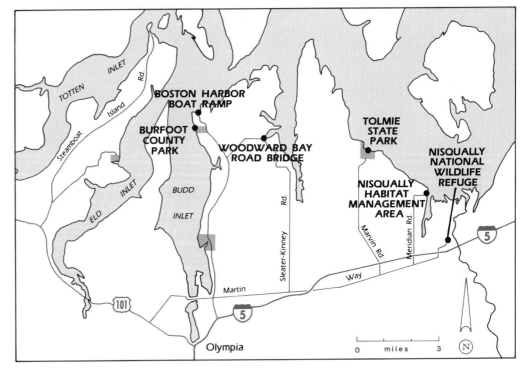

THURSTON COUNTY

	ACRES	CAMP UNITS	PICNIC UNITS	RESTROOMS	FIREPITS	SWIMMING BEACH	BOAT LAUNCH (lanes)	BOAT MOORAGE (slips/buoys)	PUBLIC PIER	DRINKING WATER	VIEWPOINT	SHORELINE LENGTH (feet)	ROCK BEACH	SAND BEACH	GRAVEL BEACH	MUD BEACH	SAND DUNES	TIDE POOLS	WETLANDS	BLUFFS
BOSTON HARBOR BOAT RAMP	1.0		•				1			•		60								
BURFOOT COUNTY PARK	50.0	25	•	•						•		1,100			•				•	•
NISQUALLY HABITAT MANAGEMENT AREA	648.3	1	•				1			•		5,280			•				•	•
NISQUALLY NATIONAL WILDLIFE REFUGE	2,820.0		•							•		39,700			•				•	•
TOLMIE STATE PARK	105.0	42	•	•	•				•	•		1,800	•	•					•	•
WOODWARD BAY ROAD BRIDGE	NA											50			•				•	

BOSTON HARBOR BOAT RAMP
Located at the community of Boston Harbor, 6 miles north of Olympia on Boston Harbor Road next to the Boston Harbor Marina (private). A small boat ramp operated by Thurston County, but developed by the State Department of Fisheries.

BURFOOT COUNTY PARK
Located on Budd Inlet six miles north of Olympia on Boston Harbor Road. The park provides restrooms, picnic areas, parking and playground equipment. A nature trail winds through a second growth coniferous forest. Trail to a gravel beach and creek estuary. Reservations can be made for group picnics. View of Budd Inlet and State Capitol Building from the beach. Interpretive trail.

NISQUALLY HABITAT MANAGEMENT AREA
The area is reached via Martin Way and Meridian Road northeast of Olympia. The area is adjacent to and operated in conjunction with the Nisqually National Wildlife Refuge. There is a paved boat launch at Luhr's Landing at the mouth of McAllister Creek. A State Game Department Conservation License is required to use this facility.

NISQUALLY NATIONAL WILDLIFE REFUGE
Located ten miles northeast of Olympia. Access is from Interstate 5 at Exit 114. Nature trails, interpretive exhibits. Visitation is by foot only to avoid disturbance of wildlife. Environmental education facilities are available for classes and organized groups. PETS NOT ALLOWED. Parts of the refuge are open to fishing during the seasons. See the separate write-up on Nisqually for more information about this site. Refuge is open during daylight hours throughout the year. Ranger on the site Mon. - Fri. Refuge is partially in Pierce County.

TOLMIE STATE PARK
Located 6 miles northeast of Olympia. Take exit 111 on Interstate 5 and follow signs 5 miles to park. Day use only. Opportunities for scuba diving to artificial underwater reefs (sunken barges), wading in relatively warm and shallow Puget Sound water. Nature trails through second growth coniferous forest. Sandy, gravelly beach. Moorage buoys available. No dock or float. Views of McNeil and Anderson islands.

WOODWARD BAY ROAD BRIDGE
Located N.E. of Olympia on Woodward Bay. Wide road shoulder at the bridge crossing allows launching of portable boats. Great for canoes and kayaks. No facilities. Limited parking on the shoulder. CAUTION: extensive soft mud at low tide. Launch and return at high tide.

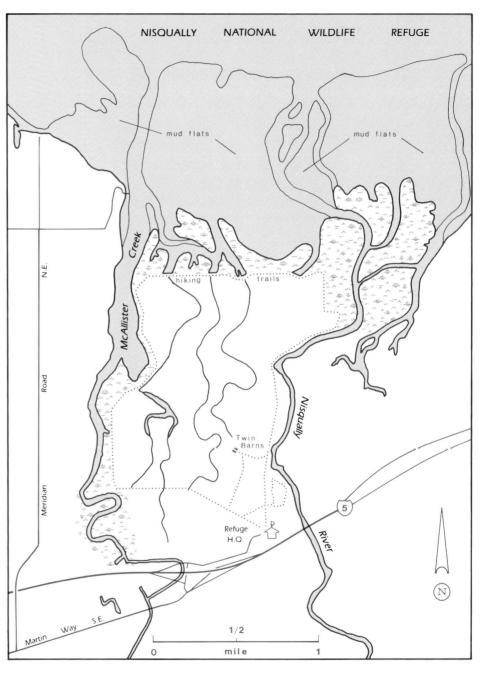

NISQUALLY NATIONAL WILDLIFE REFUGE

The Nisqually National Wildlife Refuge located on the delta of the Nisqually River in Thurston County is home for over 300 species of wildlife. The delta has an extremely diverse habitat: saltwater marshes and mudflats lie adjacent to freshwater marshes and ponds. There are also thick forests and open fields in the delta area. To protect this habitat and the wildlife that use it, the area was established as a National Wildlife Refuge in 1974.

Originally, the refuge was just a large river delta: a triangular deposit of sand and soil at the mouth of the Nisqually River. The delta was (and is) a popular fishing area for the Native Americans, and later became a legendary hunting and trapping area for white settlers. Several early Washington pioneers built homes nearby in the lush Nisqually Valley.

In 1904, Alson L. Brown purchased the land around the delta for farming. Like many delta farmers, he diked the land. With horse drawn scoops, walls were built to keep salt water out and make the rich delta land usable. In addition, wells were drilled to create a fresh water supply for the fields. It is because of these changes that the habitat of the delta is so diverse today.

By the mid-1960s farming the diked land was no longer economical and proposals were made to turn the delta into a superport or sanitary landfill. Local conservationists quickly acted to preserve the area and, as a result of their action, the National Wildlife Refuge was established.

Today, the mudflats, estuary, and saltmarshes are home for a wide variety of microorganisms as well as clams, crab, worms, insects and other invertebrates. These small organisims feed larger animals such as fish, shorebirds, herons and waterfowl. The freshwater marshes, fields, and forests at Nisqually are the home of entirely different kinds of animals. Freshwater microorganisims and invertebrates live in the marshes while mice and voles live in fields, and owls, coyotes, raccoons, weasles, and sometimes otter make their home in the forest near the river.

Nisqually is not only a haven for wildlife, but also for naturalists. Trails from 0.5 mile to 5.6 miles take hikers to the Nisqually River, through the marshes, and along saltflats. There is a special 0.5 mile long Nisqually River Interpretive Trail for those who want a self-guided tour through a riparian woodland, and there are blinds to help visitors get a closer look at wildlife

To avoid disturbance to wildlife, the refuge is open to the public only during daylight hours and access to the refuge is by foot trail only. Early morning, evening, and after a storm are good times to observe wildlife. During migration periods there is a great influx of visiting waterfowl. Spring migration usually occurs from mid-March through mid-May and fall migration from September to December. Be sure to look in a variety of habitats and along the "edges" between habitats, and remember to look high and low for wildlife. The quieter you are, the more animals you will see. Listen for animal calls and bird songs.

Binoculars or a spotting scope, as well as a field guide, are helpful for observing wildlife. You are welcome to bring a lunch and eat along the trail, but no water or trash cans are provided in the refuge. No pets or camping are allowed. Don't forget your raincoat, and remember: take only pictures, leave only footprints.

For more information contact:
Refuge Manager
Nisqually National Wildlife Refuge
100 Browns Farm Road
Olympia, WA 98506
Phone: (206)753-9467

MASON COUNTY

Mason County's 200 miles of inland shoreline includes some of the most isolated stretches of Puget Sound and Hood Canal. Where the shoreline is not developed, dense stands of conifers and madrona trees grow down to gravelly and muddy beaches. Many places along the shore have a spectacular view of the Olympic Mountains, which rise abruptly from the water's edge.

The town of Shelton, on south Puget Sound, known as "Christmas Tree Town U.S.A.," is one of the major christmas tree producing areas in the country. Shelton's waterfront on Hammersly Inlet is dominated by Simpson Timber Company's milling operations. Outside of town there are several oyster farming operations. Southeast of town is the Olympia Oyster Company; which is trying to revive the delicious and once prolific native Olympia oyster.

North of Shelton, the north bay of Case Inlet reaches to within six miles of Hood Canal. In 1962 a proposal was made to dig a channel between south Puget Sound and Hood Canal. Had it been done, it would have dramatically changed the character of the Mason county shoreline and the environment of Hood Canal.

Today, though, the shores of Hood Canal remain sparsely populated and the canal holds the reputation of being the healthiest large estuarine body of water in Washington. The oyster farms and fish markets which line the western shore of the canal are proof of the wealth of marine life there. Particularly delicious are fresh Hood Canal oysters, crab, and shrimp.

The elbow of the Hood Canal arm is the wettest area of the Puget Sound trough with rainfall averaging eighty-three inches annually. This elbow is also the home of the Skokomish Indian Reservation and Tribal Center.

Hoodsport is the major town along the canal. It is usually bustling with vacationers during the summer months, but quiet during the winter months. A state fish hatchery in town is open for visitors during working hours. Hoodsport also has a U.S. Forest Service ranger station with all the information you need on the Olympic National Forest.

North of Hoodsport, near the Mason county line, the Hamma Hamma River flows from snow capped peaks into the Canal. Hamma Hamma is an Indian word meaning "stinky stinky," which is how the river used to smell when extremely dense populations of salmon would go there to spawn, and then die. Though the salmon populations (as well as the smell) has decreased since the Indians named the river, the area still has enough salmon to make it popular for sport fishing.

The east and north shoreline of Hood Canal are particularly isolated, with some parts accessible only by a narrow dirt road which runs along a high bluff. Though it may not be possible to get to the beach, the full view of Hood Canal beneath the towering Olympic Mountain peaks is worth the trip.

MASON COUNTY
CENTER

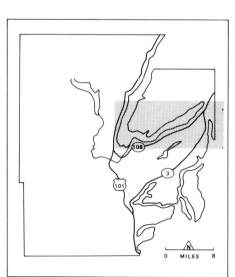

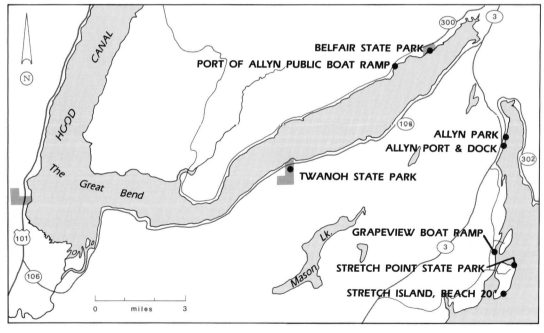

MASON COUNTY
CENTER

PUBLIC SHORE

	ACRES	CAMP UNITS	PICNIC UNITS	RESTROOMS	FIREPITS	SWIMMING BEACH	BOAT LAUNCH (lanes)	BOAT MOORAGE (slips/buoys)	PUBLIC PIER	DRINKING WATER	VIEWPOINT	SHORELINE LENGTH (feet)	ROCK BEACH	SAND BEACH	GRAVEL BEACH	MUD BEACH	SAND DUNES	TIDE POOLS	WETLANDS	BLUFFS
ALLYN PARK	2.4		•		•		1					400		•					•	
ALLYN PORT & DOCK	0.1						1		•			24	•	•					•	
BELFAIR STATE PARK	59.8	184	47	•	•	•				•		3,520		•		•				
GRAPEVIEW BOAT RAMP	0.1		•				2					30		•						
PORT OF ALLYN PUBLIC BOAT RAMP	NA						1					30	•	•						
STRETCH ISLAND, BEACH 20	NA											1,800	•	•						
STRETCH POINT STATE PARK	4.2							5				610		•						
TWANOH STATE PARK	174.7	62	111	•	•	•	3	11	•	•		2,867	•		•					

ALLYN PARK
Located 2 blocks north of Allyn City Center on Grapeview Road. Undeveloped park except for the paved boat launch lane next to Allyn Port and Dock. Low bank waterfront and open fields.

ALLYN PORT & DOCK
Located on Highway 3 north of Allyn City Center next to the post office. Single lane boat launch next to Port of Allyn Pier. Parking is available at Allyn Park. No signs on road. SHELLFISH HARVESTING NOT ALLOWED.

BELFAIR STATE PARK
Located 3 miles southwest of Belfair on Highway 300 along Hood Canal. Saltwater swimming area with sandy beaches. Large open grassy fields. Clam and oyster digging when in season.

GRAPEVIEW BOAT RAMP
Located 1 mile north of Grapeview on Grapeview Road, next door to 9's Fair Harbor Marina. Limited parking is available at the boat launch.

PORT OF ALLYN PUBLIC BOAT RAMP
Located on Hood Canal 4 miles southwest of Belfair on Northshore Road (Highway 300). Parking available near the ramp.

STRETCH ISLAND, BEACH 20
Located on the southeast side of Stretch Island which is near the town of Grapeview. Access is by boat only. Beach is composed of large boulders, cobble and gravel to a depth of 1 foot and underlain by a hard sandy clay. Only the tidelands are public.

STRETCH POINT STATE PARK
Located 12 miles north of Shelton on Stretch Island. Access is by boat only. Upland access is blocked by private ownership. Primitive, no drinking water, no facilities. Nice sandy beach.

TWANOH STATE PARK
Located on State Highway 21 nine miles west of Belfair. Large state park with marked swimming beach, wading pool for children, concession stand and tennis court. Popular boating and waterskiing area. Hiking trails, 1.25 miles and 2 miles long. Game sanctuary. Campground on opposite side of highway from the beach. Clam and oyster digging unless closed to harvest. Park hours are 6:30 a.m. until dusk.

MASON COUNTY
CENTER

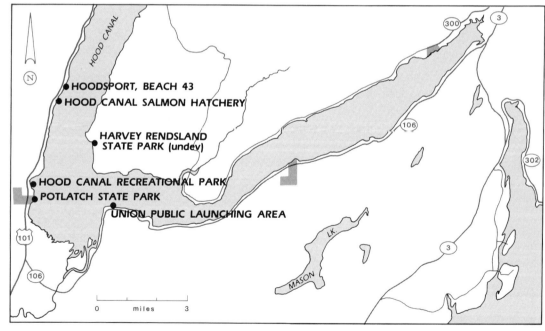

MASON COUNTY
CENTER

	ACRES	CAMP UNITS	PICNIC UNITS	RESTROOMS	FIREPITS	SWIMMING BEACH	BOAT LAUNCH (lanes)	BOAT MOORAGE (slips/buoys)	PUBLIC PIER	DRINKING WATER	VIEWPOINT	SHORELINE LENGTH (feet)	ROCK BEACH	SAND BEACH	GRAVEL BEACH	MUD BEACH	SAND DUNES	TIDE POOLS	WETLANDS	BLUFFS
HARVEY RENDSLAND STATE PARK (undev)	8.0											1,405	•		•				•	
HOOD CANAL RECREATIONAL PARK	4.7	30	•	•		2			•	•		1,000			•					•
HOOD CANAL SALMON HATCHERY	10.0		•									600			•					
HOODSPORT, BEACH 43	NA											2,951	•							
POTLATCH STATE PARK	57.0	37	82	•	•	•		5	•	•		9,570	•		•	•				
UNION PUBLIC LAUNCHING AREA	1.0			•			2					30			•					

HARVEY RENDSLAND STATE PARK (undev)
Located 2.5 miles west of the town of Tahuya along the north shore of Hood Canal, at the mouth of Rendsland Creek. No facilities. Foot traffic only. Approximately 5 acres of wetlands include a marsh and freshwater stream. Stream is dry in summer.

HOOD CANAL RECREATIONAL PARK
Located 13 miles north of Shelton on Highway 101. The park is provided by Tacoma City Light. Large parking lot. Day use only. Across from Cushman Hydroelectric Plant. Gravel bluffs in the park.

HOOD CANAL SALMON HATCHERY
Located on Hood Canal on the northside of Hoodsport on Highway 101. Department of Fisheries salmon hatchery allows visitors to walk through and look at the salmon. Access to the Beach is through the north corner of hatchery or via driveway north of the hatchery. Clam digging and oyster gathering. Lobby is open 8:00 a.m. until 4:30 p.m. weekdays. Outside area is always open.

HOODSPORT, BEACH 43
Located on Hood Canal just north of Hoodsport. Access available by boat or via car on Highway 101. Very limited parking along the highway. Cobble is the primary beach material.

POTLATCH STATE PARK
Located 12 miles north of Shelton on Highway 101. Picnic area on the beach. Open grassy areas. Good view of Hood Canal and Olympic foothills. Boat moorage buoys off beach. Good scuba diving. Clam, oysters and crab. Park hours are 6:30 a.m. until 10:00 p.m.

UNION PUBLIC LAUNCHING AREA
Take Highway 106 to Union. Launching area is in town. No facilities other than boat ramp.

MASON COUNTY
NORTH

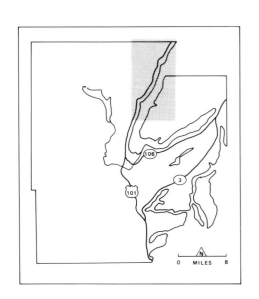

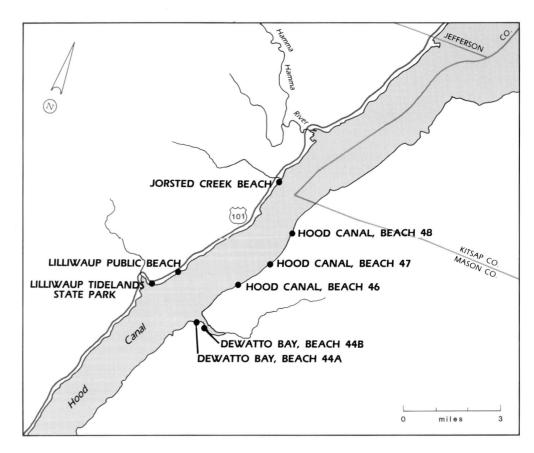

MASON COUNTY
NORTH

	ACRES	CAMP UNITS	PICNIC UNITS	RESTROOMS	FIREPITS	SWIMMING BEACH	BOAT LAUNCH (lanes)	BOAT MOORAGE (slips/buoys)	PUBLIC PIER	DRINKING WATER	VIEWPOINT	SHORELINE LENGTH (feet)	ROCK BEACH	SAND BEACH	GRAVEL BEACH	MUD BEACH	SAND DUNES	TIDE POOLS	WETLANDS	BLUFFS
DEWATTO BAY, BEACH 44A	NA											514	•	•	•					
DEWATTO BAY, BEACH 44B	NA											713	•	•	•					
HOOD CANAL, BEACH 46	NA											1,643			•					
HOOD CANAL, BEACH 47	NA											900			•					
HOOD CANAL, BEACH 48	NA											9,072			•					
JORSTED CREEK BEACH	NA											NA	•			•				
LILLIWAUP PUBLIC BEACH	NA			•								900	•		•				•	
LILLIWAUP TIDELANDS STATE PARK	42.6											4,122	•		•					•

DEWATTO BAY, BEACH 44A
Located on Dewatto Bay near Eagle Creek. Access is by boat only. Beach is composed of a gravel upper area and a sand and mud outer area. Only the tidelands are public.

DEWATTO BAY, BEACH 44B
Located on Dewatto Bay near Eagle Creek. Access is by boat only. Beach is composed of a gravel upper area and a sand and mud outer portion. Only the tidelands are public.

HOOD CANAL, BEACH 46
Located along the southern portion of Hood Canal near Eagle Creek. Access is by boat only. Beach is composed of cobble. Only the tidelands are open to the public.

HOOD CANAL, BEACH 47
Located along Hood Canal near Eagle Creek. Access is by boat only. Beach is composed of cobble. Only the tidelands are public.

HOOD CANAL, BEACH 48
Located along Hood Canal across the water from Ayock Point. Access is by boat only. Beach is composed of cobble. Only the tidelands are public.

JORSTED CREEK BEACH
Located on Hood Canal immediately south of the bridge over Jorsted Creek. Two trails lead to a flat, muddy beach which has native oysters. Adjacent beaches are private.

LILLIWAUP PUBLIC BEACH
Located 1 mile north of Lilliwaup Creek on Highway 101 along Hood Canal. Large roadside pulloff with plenty of parking. Steep trails down to a rocky beach. Beach is marked for southbound Highway 101 travelers.

LILLIWAUP TIDELANDS STATE PARK
Located on Hood Canal, 0.5 mile north of Lilliwaup on Highway 101. Unmarked gravel pullout on east side of road. Includes 1000 feet of gravel bluffs Popular location for scuba diving.

MASON COUNTY
SOUTH

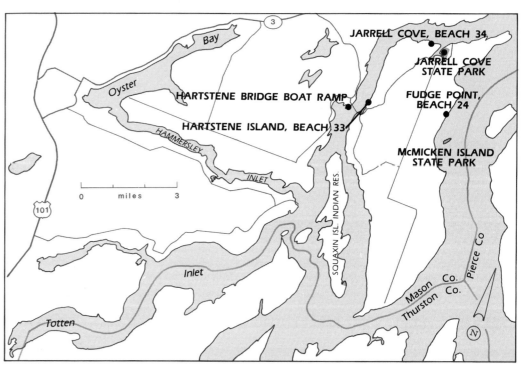

MASON COUNTY
SOUTH

	ACRES	CAMP UNITS	PICNIC UNITS	RESTROOMS	FIREPITS	SWIMMING BEACH	BOAT LAUNCH (lanes)	BOAT MOORAGE (slips/buoys)	PUBLIC PIER	DRINKING WATER	VIEWPOINT	SHORELINE LENGTH (feet)	ROCK BEACH	SAND BEACH	GRAVEL BEACH	MUD BEACH	SAND DUNES	TIDE POOLS	WETLANDS	BLUFFS
FUDGE POINT, BEACH 24	NA											5,872			•					
HARTSTENE BRIDGE BOAT RAMP	1.0			•			1					100			•					
HARTSTENE ISLAND, BEACH 33	NA											1,442	•		•					
JARRELL COVE STATE PARK	42.6	20	15	•	•			12	•	•	•	3,056			•	•			•	
JARRELL COVE, BEACH 34	NA											2,496			•	•				
McMICKEN ISLAND STATE PARK	11.5		•					5		•		1,661			•	•				

FUDGE POINT, BEACH 24
Located on the eastern side of Hartstene Island just south of Fudge Point. Access is by boat only. Beach varies from gravel to sandy gravel at extreme low tide. Only the tidelands are public.

HARTSTENE BRIDGE BOAT RAMP
Located at the west end of the Hartstene Island bridge which is over Pickering Passage. No facilities other than the ramp and parking.

HARTSTENE ISLAND, BEACH 33
Located on the western side of Hartstene Island along Pickering Passage across the water from Graham Point. Access is by boat only. There is a gravel upper beach. Only the tidelands are public.

JARRELL COVE STATE PARK
Located on Pickering Passage on the N.W. end of Hartstene Island. From Highway 3, go 8 miles on Grant Ferry Road to the park. It can also be approached from Wingert Road where it is 3.9 miles from Hartstene Bridge. Moorage fees. Game sanctuary, grassy lawns, several trails to the beach. Bulkhead and steep bank make only half of the beach usable. Shellfishing not recommended. Hours are 6:30 a.m. until 10:00 p.m. Overnight camping fee.

JARRELL COVE, BEACH 34
Located on the northern side of Hartstene Island just west of Jarrell Cove State Park. Access is by boat only. Only the tidelands are public. Beach consists of gravel with a small muddy bight in the middle of the beach at low tide.

McMICKEN ISLAND STATE PARK
Located on the east side of Hartstene Island. Access is by boat only. Primitive site with minimal facilities. Surrounded by state owned Beach 25.

MASON COUNTY
SOUTH

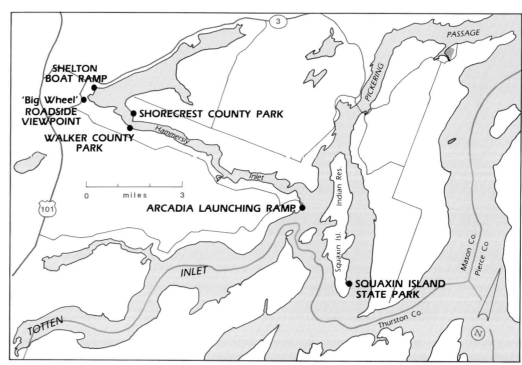

MASON COUNTY
SOUTH

	ACRES	CAMP UNITS	PICNIC UNITS	RESTROOMS	FIREPITS	SWIMMING BEACH	BOAT LAUNCH (lanes)	BOAT MOORAGE (slips/buoys)	PUBLIC PIER	DRINKING WATER	VIEWPOINT	SHORELINE LENGTH (feet)	ROCK BEACH	SAND BEACH	GRAVEL BEACH	MUD BEACH	SAND DUNES	TIDE POOLS	WETLANDS	BLUFFS
ARCADIA LAUNCHING RAMP	3.0						2					60			•					
ROADSIDE VIEWPOINT "Big Wheel"	1.0										•	NA								
SHELTON BOAT RAMP	NA						1					30			•					
SHORECREST COUNTY PARK	2.8		2	•	•	•	1			•		320			•					•
SQUAXIN ISLAND STATE PARK	31.4	20	18	•	•			28	•	•		2,673			•					
WALKER COUNTY PARK	6.5		20	•	•	•				•		1,650	•		•					•

ARCADIA LAUNCHING RAMP
From Shelton take Arcadia Road 6 miles, turn left on Arcadia Point Road and follow to boat ramp at the end of road. Parking area is 200 yards from the boat ramp.

ROADSIDE VIEWPOINT "Big Wheel"
On Highway 101 overlooking Shelton and the end of Hammersly Inlet. Interpretive signs discuss early logging history of Shelton and vicinity.

SHELTON BOAT RAMP
Located on Highway 3 in Shelton next to Simpson's Log loading facility. Parking can be a problem due to industrial activity.

SHORECREST COUNTY PARK
Take Highway 3 out of Shelton 5 miles, turn right on Agate Road and go 3.75 miles to Crestview Drive; go right 2.3 miles to Shorecrest Parkway; turn right and follow 0.5 mile to the park. The park is small with nice lawns and deciduous shade trees.

SQUAXIN ISLAND STATE PARK
Located west of Hartstene Island. Access is by boat only. Squaxin Island is an Indian Reservation; non-Indian visitors are allowed only at the park. The park has large lawn areas and beautiful views of Puget Sound islands and waters.

WALKER COUNTY PARK
Take Arcadia Road out of Shelton 1.5 miles to Walker Park Road; go 0.3 mile to park. Beach is narrow at high tide. Looks out to a narrow body of water (Hammersly Inlet). Basketball and horseshoe area. Park hours are 8:00 a.m. until 10:00 p.m. No overnight camping. Watchperson lives permanently on site.

CLAM DIGGING

Clam digging is one of the most popular recreational activities along the marine shores of Washington. Some clam digging beaches are noted in the site information in this guide and the shellfish descriptions in this book indicate where clams live on a beach. To dig clams use a shovel or tined fork to turn over a mound of beach, sort through the overturned mud and gravel, and collect the clams (hopefully!). After checking the mud for clams fill the hole back in to minimize the impact on other organisms that live in the beach.

As you collect clams, keep them cool and moist so they don't die. If you put them in a bucket of clean water two or three hours before cooking, the clams will siphon the water through their systems, cleaning out any sand or mud.

Because most clams live in the intertidal area of a beach, low tide is a good time to go digging. Different types of clams can be collected from different intertidal ranges; usually the larger the clam species, the lower it is found on the beach. Clams usually live in close proximity to one another so if you find one, dig in the same area for more.

Please respect the rights of private beachowners and dig only on public beaches. Sometimes public beaches are closed to shellfish gathering. Beaches may be closed because of paralytic shellfish poisoning (PSP, commonly known as red tide), contamination of shellfish by pollution, or due to over harvesting of the shellfish. Whatever the reason, it is important to respect closure signs for your own health and the health of shellfish populations.

There is more information on PSP in the beginning of this book or call the toll free "Red Tide Hotline" 1-800-562-5632.

SHELLFISH

Many species of shellfish live along Washington's saltwater shores. Where a particular species makes its home is generally determined by tidal influences and beach substrate. The following information will help you find and identify some of the most popular shellfish.

Hardshell Clams

Hardshell clams include Manila or Japanese clams, native littleneck clams, butter clams, and horse clams. They are found on beaches of mixed sand, gravel, and mud in protected bays. The smaller species of hardshell clams (like littlenecks) are found high on the beach and are buried shallowly, while larger clams are found low on the beach and are buried more deeply.

Manila or native littleneck clams are very similar in appearance and may reach three or four inches in length. The shells of both species have concentric rings and radiating ridges. While the native littleneck clam has a round, chalky-white shell the Manila clam has a more oblong and slightly colored shell. The Manila clam was originally imported from Japan with seed oysters and has since become well established in Puget Sound. These clams are delicious steamed and eaten with butter.

Butter clams grow to be six inches long and have chalky-white shells with concentric rings. This species may be buried as deep as ten inches, but is often much closer to the surface. The flesh of the butter clam is rubbery when steamed, but it makes excellent clam chowder.

Horse clams have a chalky-white shell that may reach up to eight inches in length. The shell is often blackened in spots by sulfides and there is usually a skin like covering, or periostracum, around the neck. The large neck, or siphon, is covered with a leather-like flap at the tip and can rarely be withdrawn all the way into the shell. Horse clams are the largest species of clam commonly found in tide flats and protected bays.

Japanese littleneck (manila) clam

Native littleneck clam

Butter clam

Horse clam

Oysters

Oysters are often found in groups attached to one another, or attached to a rock or shell. They have irregular, chalky-white shells that are often distorted to conform to the shape of the object they are attached to. Although they are found throughout the protected waters of the coast, most recreational harvesting of oysters occurs in Hood Canal. The Japanese oyster, or Pacific oyster, is by far the most common in Washington waters. Pacific oysters do not have as much flavor as the less common native oyster. During the summer oysters spawn, which makes them less desirable to eat.

Pacific oyster

Mussels

Mussels have oblong, blue-black shells and attach themselves to solid objects such as rocks or pilings by many fine hairs. The blue bay mussel of Puget Sound grows to be about three inches long, while the California mussel, which is found on the open coast, may grow to be over six inches. Mussels are easy to harvest and can be steamed or cooked like clams. They are a very popular food around the world which has only recently become popular locally.

Mussel

Cockle

Cockles are found near the surface of sand and mud beaches throughout Puget Sound. They are easily recognized by their prominent, evenly spaced ridges which fan out from the hinge area to the shell margin. The shells are light brown in color and can reach a length of about four inches. Cockles may be gathered by hand or with a garden rake and are most commonly used to make clam chowder.

Cockle

Geoduck

Geoducks (pronounced gooe-ducks) are the world's largest burrowing clam. They average about two pounds, but can weigh up to twenty pounds and their shells can reach eight inches across. The geoduck's large neck may hang out of the shell ten inches even when fully contracted. Geoducks live buried two to three feet deep. They are rare on intertidal beaches but are very abundant subtidally. On extremely low tides they can be harvested by using a large open-bottom can which prevents the sides of the hole from caving in as you dig. Usually they are harvested in deep waters by professional divers. Geoduck meat is considered a delicacy in Japan and has recently become popular in the United States.

Razor Clams

Razor clams are found on the broad, sandy, surf-swept beaches of Washington's outer coast between the Columbia River and Point Grenville. They have thin, elongated shells that may reach nine inches and are covered with a lacquer-like reddish-tan coating.

As of summer 1985 the razor clam digging season was closed because of a parasite that nearly wiped out the clam's populations. If populations revive the season will be reopened.

Softshell Clams

Softshell clams are found buried eight to fourteen inches in mud and sand bottoms near the mouths of rivers. They are abundant in Port Susan, Skagit Bay, Grays Harbor, and Willapa Bay. Their chalky-white shells may reach 6 inches in length. The shells are thin and fragile and often crack when the clam is harvested. These clams are usually collected with long-tined rakes and are very good to eat.

When harvesting clams, please respect the bag limit, season, and gear regulations that have been set up by the Department of Fisheries. These regulations are published in a booklet, "Salmon, Shellfish, Marine Fish and Sport Fishing Regulations," which is available free from the Department of Fisheries, in most marinas, and in many sporting goods stores.

Geoduck

Razor clam

Softshell clam

CLALLAM COUNTY

Clallam County, located at the "top" of the Olympic Peninsula, has 180 miles of shoreline along the Strait of Juan de Fuca and the outer Pacific coast. The county boasts the longest natural sand spit and the tallest coast Douglas-fir, and shares, with Jefferson County, the longest wilderness sand beach in the United States.

Clallam County's shoreline contains many contrasts—miles of sun-bleached drift logs, rocky headlands which flank sandy crescent beaches, feeder bluffs, sea stacks, and small offshore islands. The Pacific beaches of the county are perhaps the most unspoiled in the country. Except at La Push, they are accessible by foot trail only.

The Makah Indian reservation occupies the northernmost corner of the county. Scenic Cape Flattery, within the reservation, is the furthest northwest corner of the contiguous United States. At the tribal center in Neah Bay, the Makah Cultural and Research Center displays artifacts from the extraordinary archeological sites at Ozette and Hoko River. The museum is open daily from 10 a.m. to 5 p.m. during the summer months, and closed Monday and Tuesday from September to June. There is a modest entry fee.

Along the Strait of Juan de Fuca the coastline is mostly forested with some areas occupied by resorts and second homes. Ediz Hook, just out of Port Angeles, is a naturally formed sand spit which juts into the strait to form Port Angeles' deep-water harbor. It is worth taking the trip out Ediz Hook to get a great view of Port Angeles framed by the Olympic Mountains.

A thirty minute drive out of Port Angeles takes you to the top of Hurricane Ridge in the Olympic National Park. The ridge rises a mile above sea level and provides a spectacular view of the Olympic Mountains to the south, and the Strait of Juan de Fuca, Victoria British Columbia, and the San Juan Islands to the north.

Dungeness Spit, located between Port Angeles and Sequim, reaches five miles into the strait and is the longest natural sand spit in the country. A walk out on the spit is somewhat like taking a walk out on the ocean; the view is similar to what you would get from a boat. Designated as a National Wildlife Refuge, the spit provides ample opportunities for viewing marine birds and animals (see Dungeness National Wildlife Refuge).

Wildlife found in the coastal zone of Clallam County includes the Roosevelt elk, the rare sea otter (found near La Push), several species of whales, fur and harbor seals, the harbor porpoise, and sea lions. Bald eagles and peregrine falcons also frequent the coastal area. Migratory waterfowl are found in large concentrations at Dungeness Spit, and a substantial sea bird colony is located at Cape Flattery.

Twelve rivers of Clallam County empty from the Olympic peaks into the sea. Four species of salmon are found in Clallam County, as well as sea run and resident trout, making the area particularly popular among fishermen. Charter boats and fishing guides are available in Port Angeles and Neah Bay.

CLALLAM COUNTY
WEST

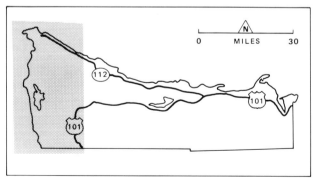

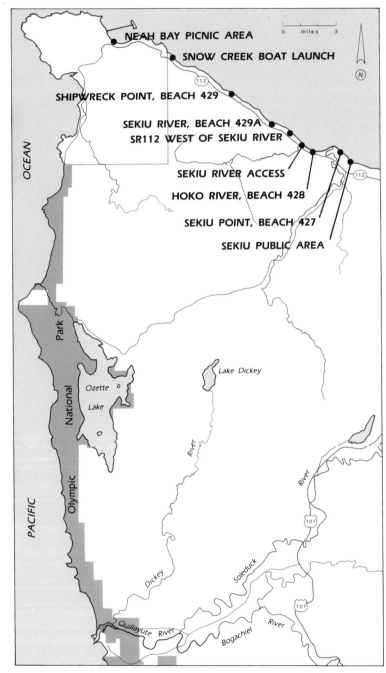

CLALLAM COUNTY
WEST

	ACRES	CAMP UNITS	PICNIC UNITS	RESTROOMS	FIREPITS	SWIMMING BEACH	BOAT LAUNCH (lanes)	BOAT MOORAGE (slips/buoys)	PUBLIC PIER	DRINKING WATER	VIEWPOINT	SHORELINE LENGTH (feet)	ROCK BEACH	SAND BEACH	GRAVEL BEACH	MUD BEACH	SAND DUNES	TIDE POOLS	WETLANDS	BLUFFS
HIGHWAY 112 WEST OF SEKIU RIVER	NA											NA	•	•						•
HOKO RIVER, BEACH 428	NA											2,750	•	•						
NEAH BAY PICNIC AREA	NA		4							•		NA								
SEKIU POINT, BEACH 427	NA											17,890	•		•					
SEKIU PUBLIC AREA	NA			•								2,500	•	•						
SEKIU RIVER ACCESS	NA											NA	•	•						
SEKIU RIVER, BEACH 429A	NA											12,210		•	•					
SHIPWRECK POINT, BEACH 429	NA											37,440	•	•						
SNOW CREEK BOAT LAUNCH	6.5	40		•								250		•						

HIGHWAY 112 WEST OF SEKIU RIVER
This is the area from Sekiu River to Bullman Creek on Highway 112 along the Strait of Juan de Fuca. Turnouts and paths to beaches 429 and 429A. Six miles of beach are accessible from this area.

HOKO RIVER, BEACH 428
Located just west of Sekiu. Access is by boat only. This very exposed beach is made up of loose sand and gravel. Only the tidelands are public.

NEAH BAY PICNIC AREA
Located in Neah Bay, on the Makah Indian Reservation. Views to Vancouver Island and North Bay Harbor. No facilities other than picnic tables.

SEKIU POINT, BEACH 427
Located between Sekiu Point and Kydaka Point west of Sekiu. Access by boat only. This very exposed beach is composed of sand and gravel. Only the tidelands are public.

SEKIU PUBLIC AREA
Located one mile west of Clallam Bay to the east of Sekiu on Highway 112. Access to the beach is readily available, park on the shoulder of the highway.

SEKIU RIVER ACCESS
Located 10 miles east of Neah Bay on Highway 112. Limited parking on the pulloffs on the side of the road. There is a sandy beach with a great view of Vancouver Island.

SEKIU RIVER, BEACH 429A
Located midway between Sekiu and Neah Bay off Highway 112. Access is available via boat or from Highway 112. CAUTION: boat landing is extremely dangerous. This exposed beach is composed of loose gravel and hardpan. Only the tidelands are public.

SHIPWRECK POINT, BEACH 429
Located just east of Neah Bay; the west boundary is distinguished by a power pole at the northeastern point of Snow Creek cove. Access by boat or via Highway 112. CAUTION: boat landing is extremely dangerous. Beach is made up of loose gravel and hardpan. Only the tidelands are public.

SNOW CREEK BOAT LAUNCH
Located 2 miles west of Neah Bay on Highway 112. Facilities on this property are operated by a concessionaire and are available for a fee.

CLALLAM COUNTY
WEST

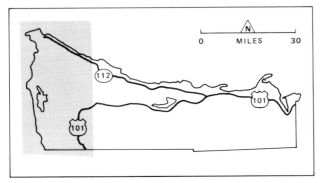

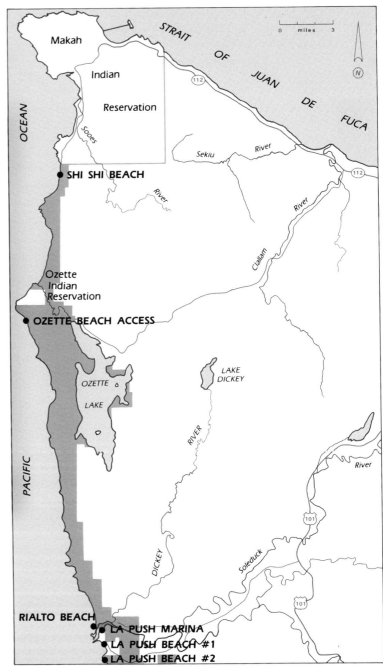

CLALLAM COUNTY
WEST

PUBLIC SHORE

	ACRES	CAMP UNITS	PICNIC UNITS	RESTROOMS	FIREPITS	SWIMMING BEACH	BOAT LAUNCH (lanes)	BOAT MOORAGE (slips/buoys)	PUBLIC PIER	DRINKING WATER	VIEWPOINT	SHORELINE LENGTH (feet)	ROCK BEACH	SAND BEACH	GRAVEL BEACH	MUD BEACH	SAND DUNES	TIDE POOLS	WETLANDS	BLUFFS
LA PUSH BEACH #1	NA											NA	•	•	•					
LA PUSH BEACH #2	NA											NA		•						
LA PUSH MARINA	NA		•				1					NA								
OZETTE BEACH ACCESS	NA		•	•								NA	•	•					•	
RIALTO BEACH	NA	15	•								•	NA	•	•						
SHI SHI BEACH	NA										•	NA	•							•

LA PUSH BEACH #1
A sandy and gravely beach in front of resorts and motels on the outskirts of La Push (Quillayute Indian fishing village). No camping. Part of the Olympic National Park coastal strip.

LA PUSH BEACH #2
Located 1 mile southeast of La Push; park on the side of the road. Three-quarter mile gravel trail through the woods to the beach. The trail leads to a boardwalk and steps down to the beach. Camping is allowed on the beach, but there are neither facilities nor fresh water. The wide sandy beach has lots of logs and sea stacks; plus beautiful rock formations.

LA PUSH MARINA
Located in La Push at the mouth of the Quillayute River. Operated by the Quillayute tribe.

OZETTE BEACH ACCESS
Take Ozette Road south from Highway 112 to Lake Ozette. The lake is 20 miles from the town of Clallam Bay. Lots of parking. No vehicle access to the beach. Ranger station. There are two trails to the beach: one is 3.0 miles long and the other 3.3 miles long. Good place for beach hikes or camping trips. The trails are boardwalks that go through woodlands and wetlands. There is also a trail to the beach from the south end of Lake Ozette (need a boat to cross the lake). Sea stacks and rocks offshore.

RIALTO BEACH
Go west on La Push Road, north of Forks. After 7 miles veer to the right on Mora Road and follow to Rialto Beach. Ample parking but no camping at the beach. Camp at Mora Campground 1 mile up river. Hike 3.0 miles up the beach to "Hole-in-Wall" rock formation. Lots of driftwood and huge logs on the beach.

SHI SHI BEACH
Located 2 miles south of Neah Bay on Cape Loop Road; turn left across the bridge, right at the "T" in the road, and follow signs 3 miles to trail head. Three-and-one-half mile trail to Shi Shi, the northern most beach of Olympic National Park. Not a very safe place to leave your car overnight.

CLALLAM COUNTY
CENTER

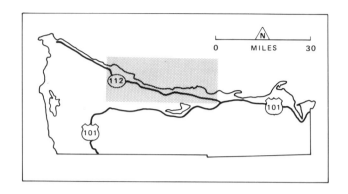

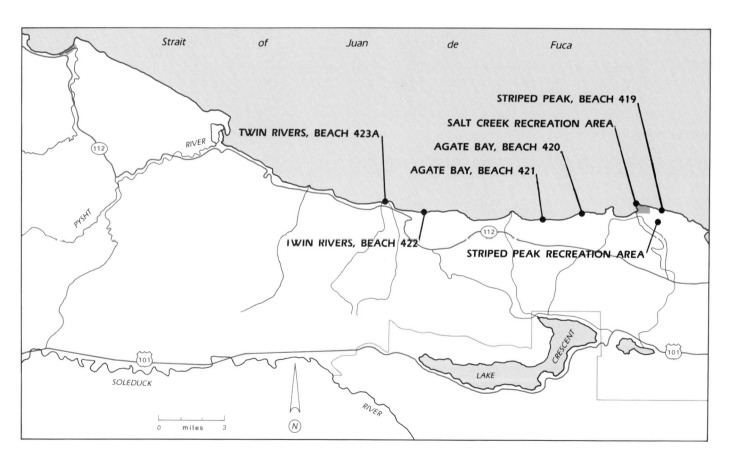

CLALLAM COUNTY
CENTER

	ACRES	CAMP UNITS	PICNIC UNITS	RESTROOMS	FIREPITS	SWIMMING BEACH	BOAT LAUNCH (lanes)	BOAT MOORAGE (slips/buoys)	PUBLIC PIER	DRINKING WATER	VIEWPOINT	SHORELINE LENGTH (feet)	ROCK BEACH	SAND BEACH	GRAVEL BEACH	MUD BEACH	SAND DUNES	TIDE POOLS	WETLANDS	BLUFFS
AGATE BAY, BEACH 420	NA											8,570	•		•					
AGATE BAY, BEACH 421	NA											8,010	•		•					
SALT CREEK RECREATION AREA	192.0	97	25	•	•	•				•		5,000	•					•	•	
STRIPED PEAK RECREATION AREA	800.0	3									•	NA								
STRIPED PEAK, BEACH 419	NA											19,570	•		•					•
TWIN RIVERS, BEACH 422	NA											30,210	•		•					
TWIN RIVERS, BEACH 423A	NA											3,415	•		•					•

AGATE BAY, BEACH 420
Located near the Salt Creek Recreation Area west of Port Angeles; the west boundary of this beach is located 900 feet east of Whiskey Creek. Access is by boat only. CAUTION: boat landing is extremely dangerous. This very exposed beach is composed of hard clay and boulders. Only the tidelands are public.

AGATE BAY, BEACH 421
Located just west of the Salt Creek Recreation Area, the west boundary of this beach is 700 feet east of Field Creek. Access is by boat only. CAUTION: boat landing is extremely dangerous. This very exposed beach is made up of hard clay and boulders. Only the tidelands are public.

SALT CREEK RECREATION AREA
From Highway 112 take Camp Hayden Road and follow signs to the recreation area. The park is at the site of historic Fort Hayden, a World War II harbor defense facility. Opportunities for hiking, beachcombing, fishing and marine life study.

STRIPED PEAK RECREATION AREA
Located west of Port Angeles. Take Lawrence Road off Highway 112 to Freshwater Bay Road; go three miles to peak (1.5 miles on gravel road). There is a scenic overlook of the Strait, Vancouver Island and the Olympics. No recreation facilities.

STRIPED PEAK, BEACH 419
Located west of Port Angeles; the west boundary of this beach is identified by distinctive vegetated rock outcropping just west of prominent Tongue Point and just east of Salt Creek drainage. The beach is composed of loose gravel, hardpan and rock bluffs. Access is by boat and hiking from both the Striped Peak Recreation Site and the Salt Creek County Area. DANGER: Boat landing is extremely hazardous.

TWIN RIVERS, BEACH 422
Located west of Salt Creek Recreation Area off of Highway 112. Access is by boat, hiking through state owned lands, or by road (PP-J-2500 off Highway 112, then take PA-J-2510 to beach). CAUTION: boat landing is extremely dangerous. This exposed beach is made up of loose gravel and hard clay.

TWIN RIVERS, BEACH 423A
Located off Highway 112 on east side of West Twin River. Parking access available from Highway 112. CAUTION: Boat landing is extremely hazardous.

CLALLAM COUNTY
CENTER

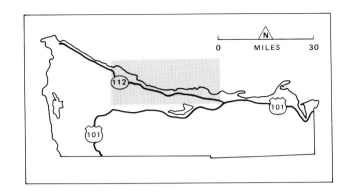

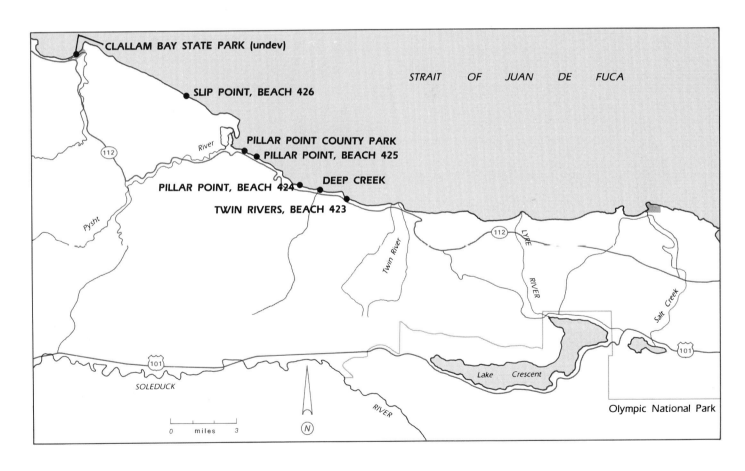

CLALLAM COUNTY
CENTER

	ACRES	CAMP UNITS	PICNIC UNITS	RESTROOMS	FIREPITS	SWIMMING BEACH	BOAT LAUNCH (lanes)	BOAT MOORAGE (slips/buoys)	PUBLIC PIER	DRINKING WATER	VIEWPOINT	SHORELINE LENGTH (feet)	ROCK BEACH	SAND BEACH	GRAVEL BEACH	MUD BEACH	SAND DUNES	TIDE POOLS	WETLANDS	BLUFFS
CLALLAM BAY STATE PARK (undev)	35.5											9,840	•	•						
DEEP CREEK	10.0	15			•							1,000	•	•						
PILLAR POINT COUNTY PARK	4.0	35	•	•		1			•			250	•		•					
PILLAR POINT, BEACH 424	NA											5,925	•	•						
PILLAR POINT, BEACH 425	NA											4,520	•	•						
SLIP POINT, BEACH 426	NA											42,750	•	•	•					•
TWIN RIVERS, BEACH 423	NA											15,365	•	•						

CLALLAM BAY STATE PARK (undev)
Located on Clallam Bay on the Strait of Juan de Fuca. Undeveloped park with no facilities.

DEEP CREEK
Unmarked pulloff on Highway 112, 2.6 miles west of the town of Twin. No facilities. Camping in trees with grassy fields. Sandy, rocky beach. This is a privately owned site where public use is allowed.

PILLAR POINT COUNTY PARK
Take Highway 112 west 1 mile east of Pysht. Protected cove view of Vancouver Island. Camping May 15 - Sept. Ranger is on the site from May 15 - Sept. 15.

PILLAR POINT, BEACH 424
Located near Silver King Resort; the west boundary of this beach is 700 feet east of Joe Creek. Access is by boat only. CAUTION: boat landing is extremely hazardous. Beach is composed of sand and loose gravel. Only the tidelands are public.

PILLAR POINT, BEACH 425
Located just east of Pillar Point County Park along the Strait of Juan de Fuca. Boating access only. CAUTION: boat landing is extremely dangerous. Beach is composed of sand, gravel and cobble. Only the tidelands are public.

SLIP POINT, BEACH 426
Located between Slip Point and Pillar Point. Access is by boat only. The exposed shoreline of this beach is made up of loose gravel and hardpan, a rock bluff and occasional boulders. Only the tidelands are public.

TWIN RIVERS, BEACH 423
Located west of Port Angeles. The west boundary of this park is 560 feet east of Deep Creek. Access is by boat or by hiking across county property from Highway 112. Limited offroad parking and no hiking trail. CAUTION: boat landing is extremely hazardous. Beach is composed of loose gravel with hard clay and boulder outcroppings.

CLALLAM COUNTY
EAST

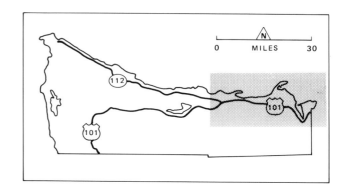

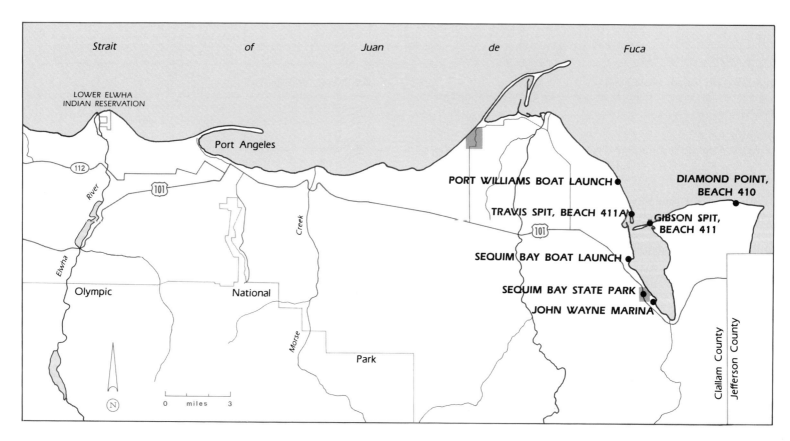

CLALLAM COUNTY
EAST

	ACRES	CAMP UNITS	PICNIC UNITS	RESTROOMS	FIREPITS	SWIMMING BEACH	BOAT LAUNCH (lanes)	BOAT MOORAGE (slips/buoys)	PUBLIC PIER	DRINKING WATER	VIEWPOINT	SHORELINE LENGTH (feet)	ROCK BEACH	SAND BEACH	GRAVEL BEACH	MUD BEACH	SAND DUNES	TIDE POOLS	WETLANDS	BLUFFS
DIAMOND POINT, BEACH 410	NA											2,710	•	•						
GIBSON SPIT, BEACH 411	NA											25,710	•	•						
JOHN WAYNE MARINA	NA						2	200				NA		•	•					
PORT WILLIAMS BOAT LAUNCH	1.0		5	•			1					500	•	•						•
SEQUIM BAY BOAT LAUNCH	39.0		5	•			3	212		•	•	2,700		•	•					
SEQUIM BAY STATE PARK	89.9	89	105	•	•		1	14		•	•	4,909		•						
TRAVIS SPIT, BEACH 411A	NA											2,800	•	•						

DIAMOND POINT, BEACH 410
Located west of Diamond Point, 1375 feet from the airport road. Access is by boat only. Beach is composed of loose gravel and sand. Only the tidelands are public.

GIBSON SPIT, BEACH 411
Located east of Sequim, the west boundary is at Travis Spit and the east boundary is at Thompson Spit. Boat access. Hiking is possible through state owned uplands, but there is no developed trail. Beach composition is loose gravel and sand.

JOHN WAYNE MARINA
Located just south of Sequim Bay State Park at Schoolhouse Point. A new development under construction in 1985.

PORT WILLIAMS BOAT LAUNCH
Follow Highway 101 to Sequim; take Brown Road north to Port Williams Road and go right for 1 mile. The area has clay bluffs.

SEQUIM BAY BOAT LAUNCH
Located on the west shore of Sequim Bay. Take West Sequim Bay Road to Pitship Point. No facilities other than the boat launch.

SEQUIM BAY STATE PARK
Located on Highway 101, 4 miles southwest of the town of Sequim. This area is especially known for its location in the rain shadow of the Olympic Mountains where it receives only 16 to 17 inches of rain annually. Activities enjoyed in the park include picnicking, camping, tennis, boating, fishing, clamming, water skiing, scuba diving, crabbing and horseshoeing. There are also several hiking trails. Also there are some of the best clay concretions found in the state. Some people call them "clay babies."

TRAVIS SPIT, BEACH 411A
Located near Sequim with the north boundary 1400 feet from Gibson Spit. Access is by boat only. Beach composition is loose gravel and sand. Only the tidelands are public.

SPITS

Spits are created when the combined wind and water currents are just right to cause suspended sand and gravel in the water to be deposited. The source of the sand and gravel is typically a river or stream, but an eroding bluff also can provide enough materials to form a spit. The deposited sediments develop into a curvilinear finger of land that reaches out into the water. By breaking wind and wave action the spit creates a quiet bay on its leeward side.

On the spit, beach grass is well adapted to the harsh environment on the exposed side, while the protected inner side allows salt marsh communities to develop. These plant communities along with driftwood are the backbone of the spit and form a first line of defense against erosion by wind and waves.

Spits play an important role in producing an environment suitable for the formation of biologically productive ecosystems. The protected bay formed by a spit is nutrient rich, and stable enough for marine vegetation and animals to become established. The bay becomes rich in algae, eelgrass, benthic (bottom dwelling) organisms, and fish. The high diversity of life in the water attracts a seemingly unending array of birds and wildlife.

Frequently these bays are feed by streams. Although streams will introduce more nutrients into the bay, they carry sediments. If the streams have a high sediment content, the bay will start to fill in. Slowly the open water will turn to a salt marsh, which may eventually form dry land.

The spit itself offers a harsh and unstable environment. Poor moisture and nutrient retention, wind and wave exposure, and changing salinities keep diversity of resident animals relatively low. The area is dominated by arthropods (crabs, insects, and spiders), that generally concentrate under driftwood and decaying algae.

Spits are of greater value to more mobile animals. They are used as resting sites by large numbers of animals, particularly birds. The isolation and unobstructed view make spits especially important for safety and freedom from disturbance. Shorebirds, gulls, terns, and waterfowl use spits as one of their primary nesting territories.

Human activity can affect spits in a variety of ways. Because they are formed by wind and wave action, spits are inherently unstable and dynamic. Any action that alters the equilibrium

of accretion and erosion can generally be counted on to change the character of the spit itself. Bulkheads, highways, or railroads along the base of nearby feeder bluffs will reduce the supply of sediments necessary to maintain a spit. Dam construction also blocks the natural flow of sediments down rivers.

On the other hand, careless logging and agricultural practices in adjacent habitats can increase river-borne sediments and smother life on the intertidal parts of spits. Increased sedimentation from streams may also hasten the filling in of the bay on the leeward side of a spit.

Recreational activities on spits have the obvious impact of frightening (or in some cases destroying) wildlife in the vicinity. If there are too many visitors, the wildlife may leave permanently. A more insidious impact is the cumulative effect of countless human feet treading on vegetation and substrate. This seemingly innocuous activity may result in the loss of stabilizing vegetation which eventually will lead to total erosion of the spit itself. Impacts on the spit environment can be moderated by avoiding vegetated portions of spits and staying clear of breeding bird colonies or otherwise sensitive areas.

CLALLAM COUNTY
EAST

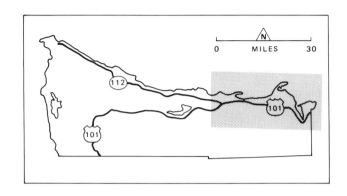

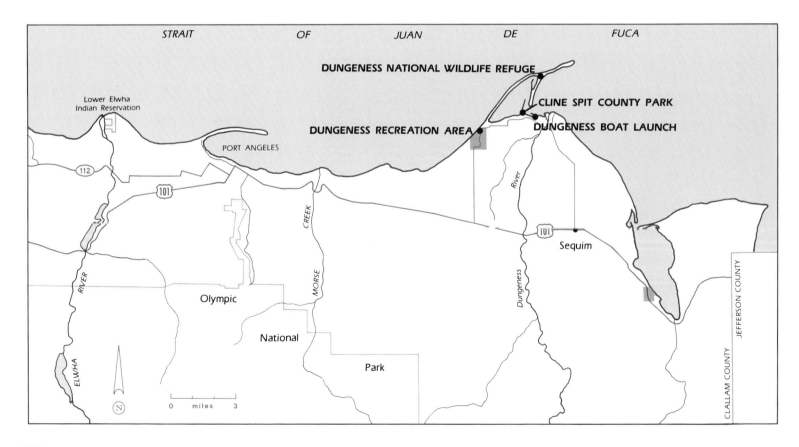

CLALLAM COUNTY
EAST

	ACRES	CAMP UNITS	PICNIC UNITS	RESTROOMS	FIREPITS	SWIMMING BEACH	BOAT LAUNCH (lanes)	BOAT MOORAGE (slips/buoys)	PUBLIC PIER	DRINKING WATER	VIEWPOINT	SHORELINE LENGTH (feet)	ROCK BEACH	SAND BEACH	GRAVEL BEACH	MUD BEACH	SAND DUNES	TIDE POOLS	WETLANDS	BLUFFS
CLINE SPIT COUNTY PARK	39.0		•				1					300		•						
DUNGENESS BOAT LAUNCH	20.0		•				2	10				1,000		•	•					
DUNGENESS NATIONAL WILDLIFE REFUGE	755.0		•								•	44,900	•	•	•					
DUNGENESS RECREATION AREA	216.0	67	20	•	•							2,500							•	

CLINE SPIT COUNTY PARK
Located west of Dungeness at a small spit. No camping allowed, but boat launching into Dungeness Bay is provided.

DUNGENESS BOAT LAUNCH
Located on Marine Drive near Cline Spit on Dungeness Bay. No facilities other than the boat launch.

DUNGENESS NATIONAL WILDLIFE REFUGE
Located 4 miles west of Sequim off of Lotzgesell Road. The Refuge is open only to foot traffic and horseback riding. Boating, fishing and clamming are activities enjoyed in the Dungeness area. Activities such as camping and making fires are allowed only in the Dungeness Recreation Area, not in the National Wildlife Refuge. The refuge is open during daylight hours throughout the year. Fall and spring are the recommended times for birders to visit the area because of the high population of migratory birds such as brant, mallard, scaup and shorebirds. Pets not allowed. There is a horse trail. Ranger on the site, weekdays.

DUNGENESS RECREATION AREA
Located on the Strait of Juan De Fuca, 3 miles north on Kitchen Road off Highway 101 between Port Angeles and Sequim. There are horse trails and hiking trails, plus access to Dungeness Spit. Campground has no electric or water hookups.

DUNGENESS NATIONAL WILDLIFE REFUGE

Dungeness Spit stretches five miles out into the Strait of Juan de Fuca. It breaks the rough sea waves of the strait to form a quiet bay, sand and gravel beaches, and tideflats where wildlife can find food and protection from wind, waves, and pounding surf. The bay and estuary of the Dungeness River produce microorganisms that form the base of a food web feeding a variety of wildlife including waterfowl, seabirds, shellfish, and anadromous and ocean fishes. Shorebirds and waterfowl feed and nest along the beaches, while seals haul out of the water to rest in the sun. The tideflats are the home of crabs, clams, oysters, and other shellfish. Shorebirds such as turnstones, phalaropes, and sandpipers search for food along the water's edge.

The refuge is open to the public during daylight hours throughout the year. The best time to observe wildlife is during fall and spring. In the fall thousands of scaup, scoters, mallards, dunlin, and plovers may be seen. Many of these stay for the winter and are joined by mergansers, cormorants, loons, and harlequin ducks. During the spring, killdeer, snipe, mallards, and pintails nest on the spit. Summer visitors will not see large numbers of birds, but may still see cormorants, great blue herons, and red-tailed hawks, as well as seals. Visitors should remember that they are guests in the homes of wildlife and should take care not to disturb them.

Dungeness National Wildlife Refuge is open to foot traffic and horseback riding only. Although there are no fires, camping, or pets allowed on the refuge, they are allowed in the Dungeness Recreation Area at the entrance to the spit.

For more information on the spit contact the Nisqually refuge office:

Refuge Manager
Nisqually National Wildlife Refuge
100 Brown Farm Road
Olympia, WA 98506
(206)753-9467

CLALLAM COUNTY
EAST

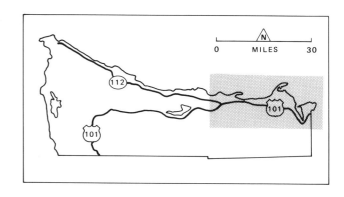
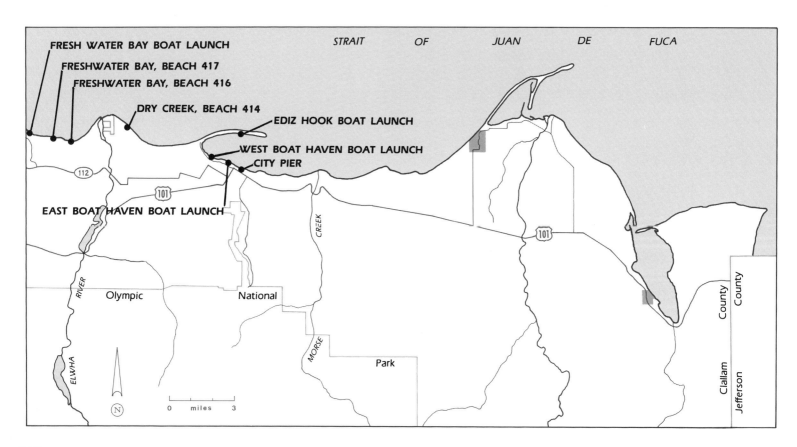

CLALLAM COUNTY
EAST

	ACRES	CAMP UNITS	PICNIC UNITS	RESTROOMS	FIREPITS	SWIMMING BEACH	BOAT LAUNCH (lanes)	BOAT MOORAGE (slips/buoys)	PUBLIC PIER	DRINKING WATER	VIEWPOINT	SHORELINE LENGTH (feet)	ROCK BEACH	SAND BEACH	GRAVEL BEACH	MUD BEACH	SAND DUNES	TIDE POOLS	WETLANDS	BLUFFS
CITY PIER	4.0		•	•				15	•	•		300	•							
DRY CREEK, BEACH 414	NA											5,580		•	•					
EAST BOAT HAVEN BOAT LAUNCH	2.0		•				1	4				60			•					
EDIZ HOOK BOAT LAUNCH	1.0			•			4	11		•		400	•	•						
FRESHWATER BAY BOAT LAUNCH	16.5		•				1					1,000	•		•	•			•	
FRESHWATER BAY, BEACH 416	NA											1,345		•	•					
FRESHWATER BAY, BEACH 417	NA											2,800		•	•					
WEST BOAT HAVEN BOAT LAUNCH	2.2		•				2	16				160			•					

CITY PIER
Located in Port Angeles one block east of the ferry terminal. Dock and pier over the water. Marine lab., open until 9:00 p.m., has many marine species on display. Covered picnic area.

DRY CREEK, BEACH 414
Located west of Port Angeles; the west boundary is 1300 feet east of Angeles Point. Access is by boat only. Beach is composed of loose gravel and hardpan. Only the tidelands are public.

EAST BOAT HAVEN BOAT LAUNCH
Follow Marine Drive 0.5 mile west of downtown Port Angeles to Port Angeles Boat Haven. Boat launch only. Restrooms are across the harbor from the boat launch.

EDIZ HOOK BOAT LAUNCH
Take Marine Drive out of Port Angeles to Ediz Hook. The beach is along the north side and the boat launch is along the south side of the "hook."

FRESHWATER BAY BOAT LAUNCH
Take Highway 112 out of Port Angeles; turn right on Lawrence Road and go 2 miles to Freshwater Bay Road; go right and drive to the end.

FRESHWATER BAY, BEACH 416
Located west of Port Angeles off of Highway 112. Access is by boat only. Beach is exposed and made up of loose gravel and hardpan. Only the tidelands are public.

FRESHWATER BAY, BEACH 417
Located west of Port Angeles; the west boundary of this beach is located 1100 feet east of Colville Creek. Access is by boat only. The composition of this exposed beach is loose gravel and hardpan. Only the tidelands are public.

WEST BOAT HAVEN BOAT LAUNCH
Take Marine Drive out of Port Angeles to the west end of the Port Angeles boat haven. Boat launch facilities only.

JEFFERSON COUNTY

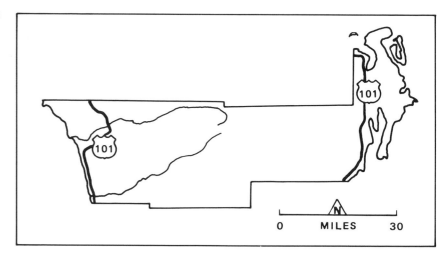

Jefferson County, stretching across the middle of the Olympic Peninsula, has shoreline totaling 207 miles. The county is unusual because it has shoreline along both the Pacific ocean, and the inland waters. The two sections of shore are divided by the high peaks of the Olympic National Park, which actually take up most of the county.

Along the outer Pacific Coast the park extends to the ocean and protects a 50-mile stretch of wilderness beach that is ideal for extended beach hiking trips. Piles of driftwood butt up to the eroding bluffs and dense forests that border the beach. These are interspersed with large rock outcrops, or sea stacks, which make these beaches unique and beautiful.

Because of the intermittent rock breaks that occur along the shore, this is an excellent area to observe tidepool life when the tide is low. The rocks also offer the perfect home for some species of birds. In addition, marine mammals can often be seen near shore; whales are occasionally seen at sea during their annual spring migration.

There is ample access to the ocean along the outer Pacific Coast. Some of the more beautiful and isolated beaches require short hikes through the woods.

On the eastern side of the county, the Olympic Mountains produce a rainshadow, creating a "banana belt" which only gets about fifteen inches of rain per year. The shoreline here winds its way along Hood Canal, Puget Sound, and the Strait of Juan de Fuca. It offers a mix of expensive resorts, historic villages, and mostly isolated gravelly beaches. A drive to the top of Mount Walker, just off highway 101, offers an excellent view of the sound, Seattle, and the Cascade and Olympic mountain ranges. A major attraction of this area is Port Townsend, a National Historic Landmark.

Once a major seaport, Port Townsend is now considered the best example of a Victorian seacoast town north of San Francisco. The downtown section has a wide variety of restaurants and many interesting shops. Nearby neighborhoods are dotted with elaborate Victorian homes. Some are open to the public during annual home tours and several have been converted into bed-and-breakfast establishments. There are also several art galleries, a museum, and festivals for different occasions throughout the year, one of the most famous of which is the Wooden Boat Festival in late summer.

Fort Flagler and Fort Worden, were built in northern Jefferson county during the late nineteenth century to provide Puget Sound cities with protection against naval attacks. Old Fort Townsend, which is of an earlier era, is also nearby. When the forts became obsolete they were purchased by Washington State. They are now large state parks where the historic buildings are open to visitors. The forts are all within fifteen miles of Port Townsend. (See Forts write-up.)

Protection Island, east of Port Townsend in the Straight of Juan de Fuca, is in the process of being purchased by the U.S. Fish and Wildlife Service and will be converted into a National Wildlife Refuge. The island has the fourth largest nesting population of rhinoceros auklets in the world and boasts to be the nesting site of 70 percent of breeding seabirds in Puget Sound. Some seabirds that are found there other than the rhinoceros auklet include glaucous winged gulls, tufted puffins, pigeon guillemot, oystercatchers, and cormorants. There is currently no public access to Protection Island.

BEACH HIKING

Some of the last wilderness beaches in the contiguous United States are found on the outer coast of the Olympic Peninsula. Because many of these beaches are accessible only by foot they have become increasingly popular for wilderness beach hiking. Hikers are rewarded with scenic cliffs, rocky headlands, islands, tidepools and seastacks.

The three most popular areas for beach hikes are the stretches of beach between the Hoh River and the town of La Push; north of La Push to the Ozette Ranger Station; and from the Ozette ranger station to Shi Shi beach just south of the Makah Indian reservation.

While traveling on the beach, it is important to carry a tide table and be aware of the tides. Traveling is easiest at low tide and can be very unsafe at high tide. Even an average high tide will cover most of the beach and make hiking in some areas impossible. It is particularly important to avoid attempting to round major headlands when the tide is coming in. Many sections of the beach have a trail in the woods over the headlands so you can continue traveling at high tide. Also, be cautious of getting stranded on rocky bluffs that are easily accessible at low tide, but may be a cold swim at high tide.

Typically the weather along the outer coast is windy and wet. Before you take off, find out if there are any major storms approaching by checking in with the ranger station, or by monitoring the NOAA Weather advisory on 162.55MHz. Even if the forecast is clear it is important to be prepared for anything: bring extra wool or pile clothing and good rain gear. During a storm beware of drift logs getting tossed up on the beach, and streams that may crest from rainfall and be impossible to cross.

The National Park Service recommends purification of all water, particularly at more heavily used campsites, to eliminate any threat of contamination. Water may be purified by boiling for 60 seconds, by adding chlorine or iodine drops (available at outdoor stores), or by carrying a small water filter.

Black bears have been a recurring problem along the coast. To keep bears out of your food, store it out of reach. If a bear is allowed to get a free meal once, it will continue to come back and harass campers. Also look out for mice, chipmunks, ravens, raccoons, and skunks all of whom will help themselves to any food not stored adequately.

In order to preserve the wilderness environment, the Park Service has set up some guidelines for visitors to the coast:

1. Pets, vehicles and weapons are prohibited on trails to avoid disturbing wildlife, and to ensure a wilderness experience for visitors.

2. Keep the maximum group size less than 12. Large groups disturb wildlife, damage trails and campsites more, and detract from the experience of others.

3. Do not build fires in piles of driftwood where the fire may spread into a large beach fire. Driftwood can be excellent fuel and is a renewable resource on the outer coast, but it should be used in an existing fire ring or in the sand below the tide line where fire scars will be erased.

4. Pack out all trash, except for paper which can be completely burned. Do not throw trash down outhouse holes.

5. Prevent pollution of streams by burying all human waste at least 100 feet away from any running water. Also keep all soap away from water sources and do not wash dishes in streams.

6. Fill out a back country permit for overnight camping and also leave word with a responsible party of your intended route and estimated time of return. If there is an emergency, notify either the Lake Ozette Ranger Station (206-963-2725), the Mora Ranger Station (206-374-5460), or the 24-hour emergency number (206-452-9545).

If you are aware of and conscientious about avoiding the environmental hazards involved with wilderness beach hiking your trip along the outer coast can be a rare and wonderful adventure.

JEFFERSON COUNTY
WEST

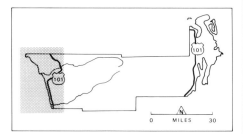

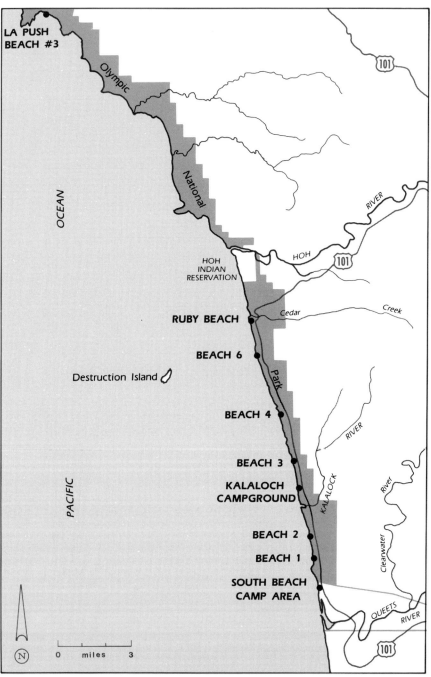

JEFFERSON COUNTY
WEST

	ACRES	CAMP UNITS	PICNIC UNITS	RESTROOMS	FIREPITS	SWIMMING BEACH	BOAT LAUNCH (lanes)	BOAT MOORAGE (slips/buoys)	PUBLIC PIER	DRINKING WATER	VIEWPOINT	SHORELINE LENGTH (feet)	ROCK BEACH	SAND BEACH	GRAVEL BEACH	MUD BEACH	SAND DUNES	TIDE POOLS	WETLANDS	BLUFFS
BEACH 1 (Olympic National Park)	NA			•								NA	•	•						
BEACH 2 (Olympic National Park)	NA			•								NA	•	•						
BEACH 3 (Olympic National Park)	NA											NA	•	•						•
BEACH 4 (Olympic National Park)	NA			•								NA	•	•						
BEACH 6 (Olympic National Park)	NA			•								NA	•	•						•
KALALOCH CAMPGROUND (Olympic National Park)	101.3	195	15	•	•					•		100	•	•						
LA PUSH BEACH #3	NA			•								NA	•	•						
RUBY BEACH (Olympic National Park)	NA			•							•	NA		•					•	
SOUTH BEACH CAMP AREA (O.N.P.)	NA			•								NA	•	•						

BEACH 1 (Olympic National Park)
Located 2 miles south of Kalaloch on Highway 101. Trail to the beach is through the woods. Limited parking. Pit toilet. This site is part of the Olympic National Park coastal strip.

BEACH 2 (Olympic National Park)
Located 1 mile south of Kalaloch on Highway 101. Limited parking. Pit toilet. Short trail to the beach. This site is part of the Olympic National Park coastal strip.

BEACH 3 (Olympic National Park)
Located 1 mile north of Kalaloch on Highway 101. There are several roadside pulloffs along this section of the beach. Good views of sea stacks. No facilities. Limited parking. Short trails down the bluffs to the beach. This site is part of the Olympic National Park coastal strip.

BEACH 4 (Olympic National Park)
Located 3 miles north of Kalaloch on Highway 101. Large parking area with trails to the beach. No facilities or water. View of Destruction Island. This site is part of the Olympic National Park coastal strip.

BEACH 6 (Olympic National Park)
Located 2 miles south of Ruby Beach on Highway 101. Gravel parking lot. Good whale watching area, particularly for seeing migrating California gray whales. View of Destruction Island. No camping. This site is part of the Olympic National Park coastal strip.

KALALOCH CAMPGROUND (Olympic National Park)
Located 1 mile north of Kalaloch. This is a large site with lots of camping on the bluff above the beach. 1 mile loop trail. The interpretive center south of the campground operates learning programs and is open during the summer. Recreational vehicle holding tank dump station.

LA PUSH BEACH #3
Located approximately 2 miles southeast of La Push; park on the side of the road. 1.3 mile gravel trail to the beach. Camping on the beach. No facilities or water.

RUBY BEACH (Olympic National Park)
Located 8 miles north of Kalaloch on Highway 101. Scenic overlook. Freshwater creek. No camping. Trail to the beach.

SOUTH BEACH CAMP AREA (O.N.P.)
Located along Highway 101 just north of the Queets River bridge. Overflow area for developed campgrounds further north. No camp facilities. R.V. and auto parking available. Tents could be put up in gravel parking lot. Open for camping in the summer only. No water available.

RAIN FORESTS

The Humid-Temperate weather zone of the Pacific coast is an ideal climate to produce a lush coniferous rain forest. The Hoh and Quinault River areas of the Olympic National Forest and Olympic National Park are well known for their rain forests. Precipitation in excess of 140 inches per year in this area produces forests of Sitka spruce, western hemlock and western red cedar that reach gigantic proportions.

The roof of the rain forest is a canopy of tree crowns, and the light that filters through has an emerald-green hue that seems to radiate from leaves and needles. Mosses hang like gray-green tapestries from tree limbs, and the forest is covered with a green carpeting of moss, sword fern, lady fern, deer fern, oxalis, and wild lily of the valley. Between the floor and roof of the rain forest, a variety of smaller shrubs and bushes grow between the huge tree trunks. Particularly adapted to this existence is the coast rhododendron, salal, and huckleberry.

The life cycle of the forest is a perpetual exchange of birth and death. In this lush community forest trees may grow to mature adults reaching 200 feet in height high and 60 feet in girth at the base. When these gigantic trees finally die and fall to the forest floor, a breeding ground is created on the fallen logs for mosses, fungi, lichens, and young hemlock which help decompose the log into rich humus and soil.

Often, new hemlock trees are born on the fallen logs as seeds germinate and send roots through or around the old nurse log. After the old log decays, the tree that germinated on it appears as though standing on legs because the roots were formed above ground level.

The rain forest provides a haven for small and large animals. Among the more common are deer mice, western pine squirrels, coyotes, cougars, bobcats, beavers, black-tailed deer, Roosevelt elk, and a variety of birds.

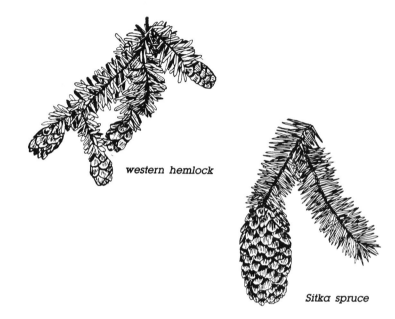

western hemlock

Sitka spruce

western red cedar

JEFFERSON COUNTY
NORTH

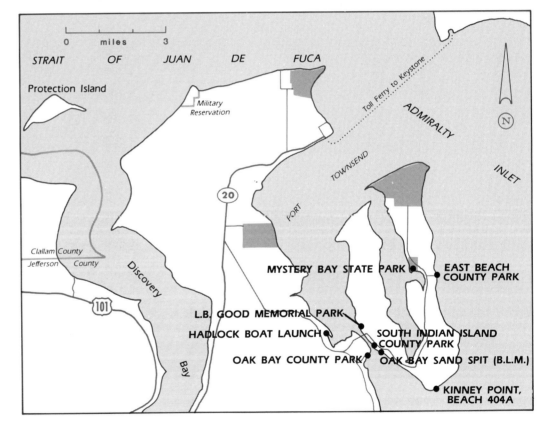
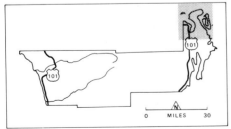

JEFFERSON COUNTY
NORTH

PUBLIC SHORE

	ACRES	CAMP UNITS	PICNIC UNITS	RESTROOMS	FIREPITS	SWIMMING BEACH	BOAT LAUNCH (lanes)	BOAT MOORAGE (slips/buoys)	PUBLIC PIER	DRINKING WATER	VIEWPOINT	SHORELINE LENGTH (feet)	ROCK BEACH	SAND BEACH	GRAVEL BEACH	MUD BEACH	SAND DUNES	TIDE POOLS	WETLANDS	BLUFFS
EAST BEACH COUNTY PARK	8.1		2	•	•	•					•	100		•						
HADLOCK BOAT LAUNCH	NA			•			1			•		NA								
KINNEY POINT, BEACH 404A	NA											3,900	•	•						
L.B. GOOD MEMORIAL PARK	NA		8		•							NA	•		•					
MYSTERY BAY STATE PARK	2.8		5	•			1	27	•	•		100		•	•					
OAK BAY COUNTY PARK	31.2	20	7	•	•		1					2,040		•					•	
OAK BAY SAND SPIT (B.L.M.)	5.0											2,500	•	•	•					
SOUTH INDIAN ISLAND COUNTY PARK	53.7		7	•					•			11,350	•	•	•			•		

EAST BEACH COUNTY PARK
Located on the east shore of Marrowstone Island. The park is reached by taking East Beach Road off Fort Flagler Road. A small mostly undeveloped park. Nice picnic shelter and a view of Mt. Baker.

HADLOCK BOAT LAUNCH
Out of Hadlock City Center take Oak Bay Road to Lower Hadlock. This is a boat launch only.

KINNEY POINT, BEACH 404A
Located at the southern end of Marrowstone Island. Access is by boat only. The upper beach is made up of sand, the lower beach composed of sand and gravel. Only the tidelands are public.

L.B. GOOD MEMORIAL PARK
Located on Indian Island. This is a small county park near Indian Island Bridge. Grassy area leading to portage canal. Minimal development. Provides access to state owned tidelands.

MYSTERY BAY STATE PARK
Located on Fort Flagler Road on the west side of Marrowstone Island. There is a pier with moorage spaces.

OAK BAY COUNTY PARK
Located on Oak Bay Road 1.0 mile south of Hadlock. There are two park access roads. Camping is on the spit or on grassy uplands. Park has wetlands.

OAK BAY SAND SPIT (B.L.M.)
This is the sand spit at the head of Oak Bay, on the south shore of Indian Island. Surrounded by state owned tidelands.

SOUTH INDIAN ISLAND COUNTY PARK
Located on South Indian Island 0.5 mile past the bridge on the west side of Indian Island. Minimal development. View of Mt. Rainier and the Olympics.

JEFFERSON COUNTY
NORTH

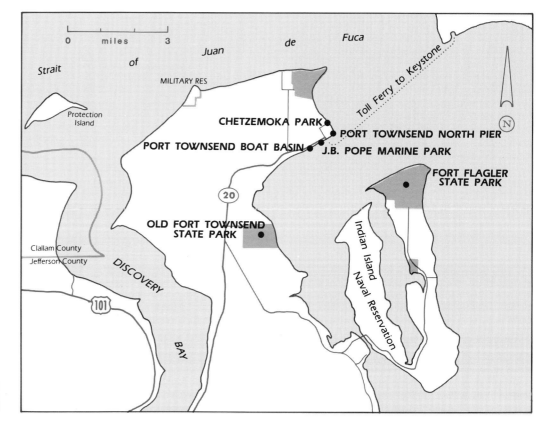
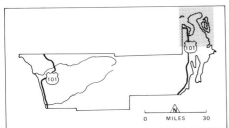

JEFFERSON COUNTY
NORTH

	ACRES	CAMP UNITS	PICNIC UNITS	RESTROOMS	FIREPITS	SWIMMING BEACH	BOAT LAUNCH (lanes)	BOAT MOORAGE (slips/buoys)	PUBLIC PIER	DRINKING WATER	VIEWPOINT	SHORELINE LENGTH (feet)	ROCK BEACH	SAND BEACH	GRAVEL BEACH	MUD BEACH	SAND DUNES	TIDE POOLS	WETLANDS	BLUFFS
CHETZEMOKA PARK	10.1		9	•	•	•				•	•	750	•	•				•		•
FORT FLAGLER STATE PARK	793.4	154	40	•	•		2	15		•	•	30,920	•	•	•				•	•
J.B. POPE MARINE PARK	1.2		5	•		•			•	•	•	3,600		•						
OLD FORT TOWNSEND STATE PARK	376.7	52	64	•	•					•		3,200	•	•						•
PORT TOWNSEND BOAT BASIN	NA		2	•			1	450	•			NA								
PORT TOWNSEND NORTH PIER	NA						2		•	•	•	NA								

CHETZEMOKA PARK
Located at 950 Jackson Street on Admiralty Inlet in Port Townsend. Small waterfront day use park.

FORT FLAGLER STATE PARK
Located on the north end of Marrowstone Island. The park contains a variety of environments which include bluffs and wetlands. Other features are a nature trail, interpretive historical exhibit, gun batteries and a marsh. Scuba diving is allowed.

J.B. POPE MARINE PARK
Located in the northeast end of downtown Port Townsend, across the street from the historic museum. Includes Marine Park Community Building plus a dock and pier. Good scuba diving offshore.

OLD FORT TOWNSEND STATE PARK
Located 5.0 miles south of Port Townsend on Highway 113. Turn on Old Fort Townsend road. This is the site of an early fort which was built for defense against hostile indians. Interpretive exhibits, open grassy field (old parade ground) and a trail to the beach. View of Port Townsend and the North Cascade mountains.

PORT TOWNSEND BOAT BASIN
Located on Highway 113 just before Port Townsend on the right. There are full boating facilities and a jetty. Customs check.

PORT TOWNSEND NORTH PIER
Located at the northeast end of downtown Port Townsend. No facilities other than the pier. Great views.

JEFFERSON COUNTY
NORTH

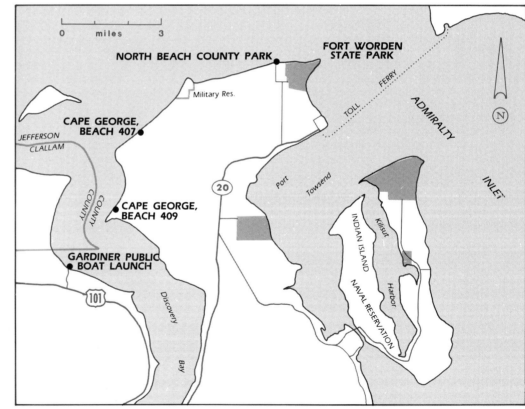
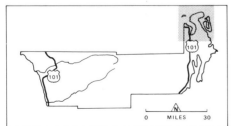

JEFFERSON COUNTY
NORTH

	ACRES	CAMP UNITS	PICNIC UNITS	RESTROOMS	FIREPITS	SWIMMING BEACH	BOAT LAUNCH (lanes)	BOAT MOORAGE (slips/buoys)	PUBLIC PIER	DRINKING WATER	VIEWPOINT	SHORELINE LENGTH (feet)	ROCK BEACH	SAND BEACH	GRAVEL BEACH	MUD BEACH	SAND DUNES	TIDE POOLS	WETLANDS	BLUFFS
CAPE GEORGE, BEACH 407	NA											5,035	•	•	•					
CAPE GEORGE, BEACH 409	NA											1,475	•	•						
FORT WORDEN STATE PARK	340.6	52	43	•	•		2	12		•	•	9,590		•			•			
GARDINER PUBLIC BOAT LAUNCH	NA						1					NA								
NORTH BEACH COUNTY PARK	0.9		4	•	•	•					•	310	•						•	

CAPE GEORGE, BEACH 407
Located near the Cape George Military Reservation on Port Discovery Bay, 2000 feet east of the marina at Cape George. Access is by boat only. The beach is sand, cobble and hardpan. Only the tidelands are public.

CAPE GEORGE, BEACH 409
Located on Port Discovery Bay, this beach is 4000 feet northeast of the pond at Beckett Point, and 8180 feet southwest of Beach 407. Access is by boat only. Composition of the beach is sand, gravel and cobble. Only the tidelands are public.

FORT WORDEN STATE PARK
Located 1.5 miles north of Port Townsend; take Cherry Street north out of town. The park overlooks the Strait of Juan de Fuca. This area provides a variety of recreational opportunities such as an underwater park for divers, a nature marine trail, biking and hiking trails. Other habitats to be explored include dunes, tidepools (at low tide) and lots of sandy beach which is both exposed and protected. The coast guard is located at the point. The park also has a collection of Victorian houses, barracks and other army structures. Fort Worden is designated as a Youth Hostel.

GARDINER PUBLIC BOAT LAUNCH
Located just north of Gardiner on Port Discovery Bay. No facilities other than the boat launch.

NORTH BEACH COUNTY PARK
Located north of Port Townsend. Take Highway 113 to San Juan Ave.; go left on 49th street, right on Kuhn Street and follow to the end. A quiet park in a residential area. Lawns and a great view of Mt. Baker, ships, and Vancouver Island to the north.

JEFFERSON COUNTY
CENTER

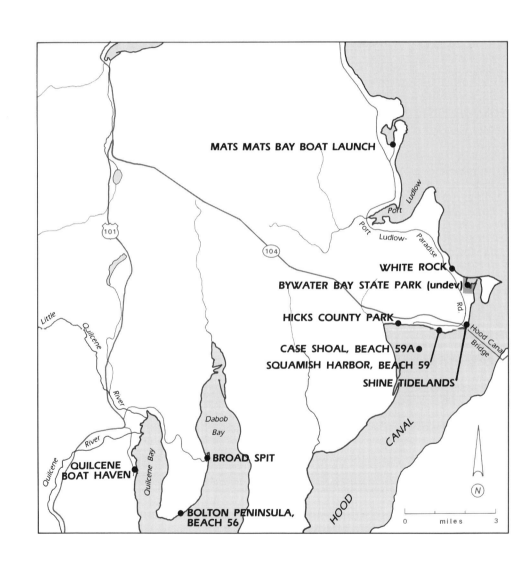
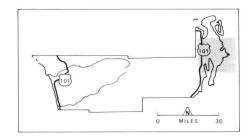

JEFFERSON COUNTY
CENTER

PUBLIC SHORE

	ACRES	CAMP UNITS	PICNIC UNITS	RESTROOMS	FIREPITS	SWIMMING BEACH	BOAT LAUNCH (lanes)	BOAT MOORAGE (slips/buoys)	PUBLIC PIER	DRINKING WATER	VIEWPOINT	SHORELINE LENGTH (feet)	ROCK BEACH	SAND BEACH	GRAVEL BEACH	MUD BEACH	SAND DUNES	TIDE POOLS	WETLANDS	BLUFFS
BOLTON PENINSULA, BEACH 56	NA											2,400	•							
BROAD SPIT	2.0											1,000		•		•				
BYWATER BAY STATE PARK (undev)	134.6											21,944	•	•						
CASE SHOAL, BEACH 59A	NA											NA			•					
HICKS COUNTY PARK	0.7		1	•			1				•	460	•							
MATS MATS BAY BOAT LAUNCH	NA		•				1		•			40			•					
QUILCENE BOAT HAVEN	NA	2	•		•		2	30				100			•					
SHINE TIDELANDS	NA	20		•			3					1,500			•					
SQUAMISH HARBOR, BEACH 59	NA											1,335		•	•					
WHITE ROCK	71.0											1,500			•	•				

BOLTON PENINSULA, BEACH 56
Located at the southern end of the Bolton Peninsula on Hood Canal. Access is by boat only. Upper beach is cobble. Lower beach is gravel, cobble and scattered boulders. Only the tidelands are public.

BROAD SPIT
Located 3.0 miles east of Quilcene. A sand spit on the east shore of Bolton Peninsula. Tideland with a small tidal flat behind the spit. Surrounded by private land. Access is by boat only.

BYWATER BAY STATE PARK (undev)
Located 1.0 mile north of the Hood Canal Bridge along west side of Bywater Bay.

CASE SHOAL, BEACH 59A
Located in Squamish Harbor south of Shine. Access is by boat only. Beach is gravel.

HICKS COUNTY PARK
Located on Hood Canal near Shine. This is a small park with a great view of the Olympics and Hood Canal.

MATS MATS BAY BOAT LAUNCH
Located at the south end of Mats Mats Bay which is 2.0 miles north of Port Ludlow.

QUILCENE BOAT HAVEN
From Quilcene take Rogers Street to Linger Longer Road; turn left and go 1.5 miles. Small protected harbor for moorage. Jetty to walk on.

SHINE TIDELANDS
Take first road to north directly after West Hood Canal Bridge on Highway 104; take Paradise Bay Road and then go right on Terminator Point Road. There is also access 0.5 mile down Paradise Bay Road; go right on Seven Sisters Road and 0.5 mile to parking and gravel launch. No camping. Mostly undeveloped area next to bridge. Entire area for 1.0 mile north along the beach is managed by the State Department of Natural Resources.

SQUAMISH HARBOR, BEACH 59
Located in Squamish Harbor along Hood Canal just east of Shine. Access is by boat only. Only the tidelands are public. There is a broad gravel upper beach and a flat sand lower beach.

WHITE ROCK
Located 3.0 miles southeast of Port Ludlow on Hood Canal. Some mud flats and tidelands plus a small creek. Access via county road.

JEFFERSON COUNTY
SOUTH

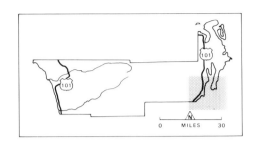

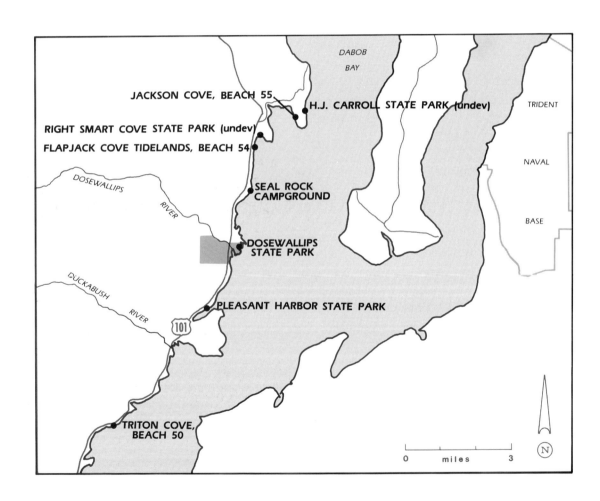

JEFFERSON COUNTY
SOUTH

PUBLIC SHORE

	ACRES	CAMP UNITS	PICNIC UNITS	RESTROOMS	FIREPITS	SWIMMING BEACH	BOAT LAUNCH (lanes)	BOAT MOORAGE (slips/buoys)	PUBLIC PIER	DRINKING WATER	VIEWPOINT	SHORELINE LENGTH (feet)	ROCK BEACH	SAND BEACH	GRAVEL BEACH	MUD BEACH	SAND DUNES	TIDE POOLS	WETLANDS	BLUFFS
DOSEWALLIPS STATE PARK	424.5	155	35	•	•							5,250		•	•				•	
FLAPJACK COVE TIDELANDS, BEACH 54	NA											567			•					
H.J. CARROLL STATE PARK (undev)	2.8											560			•					
JACKSON COVE, BEACH 55	NA											2,791	•		•					
PLEASANT HARBOR STATE PARK	0.8		•					10				100	•		•					
RIGHT SMART COVE STATE PARK (undev)	2.8											100	•		•					
SEAL ROCK CAMPGROUND	30.0	35	10	•	•					•	•	2,700	•		•	•		•		
TRITON COVE, BEACH 50	NA											2,610	•	•						

DOSEWALLIPS STATE PARK
Located north of Brinnon on Highway 101. The park provides access to Dosewallips River. There is a trail through the wetlands to the beach. Camping on opposite side of Highway 101 from the water. Grassy picnic area. View of Mt. Rainier. Clam digging allowed.

FLAPJACK COVE TIDELANDS, BEACH 54
Located along Hood Canal just south of Jackson Cove. The beach can be reached from Highway 101.

H.J. CARROLL STATE PARK (undev)
Located north of Brinnon on Pulali Point. Park is not marked and has no facilities for recreation.

JACKSON COVE, BEACH 55
Located in Jackson Cove adjacent to Pulali Point. Access is by boat only. The north half of the beach is composed of gravel and the south half is rock ledge. All of the surrounding beach and uplands are privately owned.

PLEASANT HARBOR STATE PARK
Located 1.5 miles north of Duckabush on Highway 101. Take the single lane curvy road which is next to the white buildings with the Pleasant Harbor sign. A very limited development with a rocky, gravely beach. Moorage on the state park float which is north of private moorage. Signs for the park can be seen from the water only.

RIGHT SMART COVE STATE PARK (undev)
Take Highway 101 three miles north of Dosewallips to the boat launch. This property is a small undeveloped plot east of the boat launch.

SEAL ROCK CAMPGROUND
Located north of Brinnon on Highway 101. Under private operator lease from the national forest. Fees for day use as well as camping. Camping is in the trees. Oyster beach; season is open from May 15 - July 15. State laws apply.

TRITON COVE, BEACH 50
Located just south of McDaniel Cove along Hood Canal. Access is by boat only. The beach is composed of cobble, except the southern half has sand. Only the tidelands are public.

JEFFERSON COUNTY
SOUTH

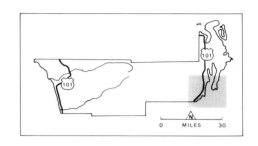

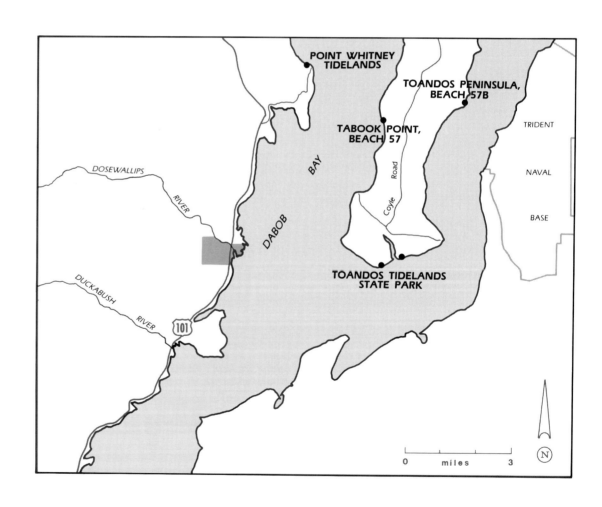

JEFFERSON COUNTY
SOUTH

	ACRES	CAMP UNITS	PICNIC UNITS	RESTROOMS	FIREPITS	SWIMMING BEACH	BOAT LAUNCH (lanes)	BOAT MOORAGE (slips/buoys)	PUBLIC PIER	DRINKING WATER	VIEWPOINT	SHORELINE LENGTH (feet)	ROCK BEACH	SAND BEACH	GRAVEL BEACH	MUD BEACH	SAND DUNES	TIDE POOLS	WETLANDS	BLUFFS
POINT WHITNEY TIDELANDS	10.0		•		•		1		•	•		2,000		•						
TABOOK POINT, BEACH 57	NA											3,280	•	•						
TOANDOS PENINSULA, BEACH 57B	NA											12,050	•	•						
TOANDOS TIDELANDS STATE PARK	0.8											10,455	•	•						

POINT WHITNEY TIDELANDS
Take Bee Mill Road off Highway 101 to Pt. Whitney Road. This is a Department of Fisheries shellfish research station. Display of marine species. Boat launch to Dabob Bay. Pier available for fishing. The lagoon is also owned by the Department of Fisheries and is open for clam and oyster harvest on an irregular basis.

TABOOK POINT, BEACH 57
Located on the western side of Toandos Peninsula just south of Tabook Point. Access is by boat only. Beach has a cobble, gravel upper area and a broad, sandy lower area. Only the tidelands are public.

TOANDOS PENINSULA, BEACH 57B
Located on the eastern side of Toandos Peninsula fronting the naval reservation. Access is by boat only. Beach is composed of large cobble in the upper area and sand in the lower area. Only the tidelands are state owned.

TOANDOS TIDELANDS STATE PARK
Located at the tip of Toandos Tidelands (S.W. Section). It is the section of land located on both sides of Fisherman's Harbor. Accessible by boat only.

GRAYS HARBOR COUNTY

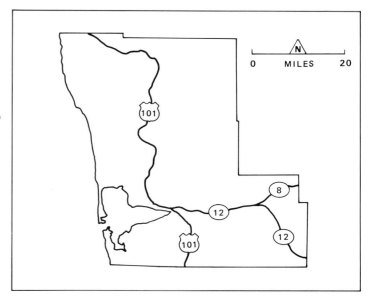

Grays Harbor County has fifty-seven miles of exposed shoreline along the outer coast that is broken by a large estuary, Grays Harbor, which has eighty-nine miles of protected saltwater shoreline. The harbor, named for the eighteenth-century American naval explorer Robert Gray, is a large estuary fed by five rivers, the largest of which is the Chehalis. The isolated outer coast contrasts with the shore of the Grays Harbor which is largely dominated by the industrial areas of Aberdeen and Hoquiam.

Aberdeen and Hoquiam, the county's two major cities, reflect the life force of the area, which is lumber. 74% of all land in the county south of the Olympic National Park is owned by private individuals and companies in the lumber business. Milling and shipping operations along Grays Harbor's shore process the prolific Douglas-fir that are harvested in the surrounding area. Many of these lumber companies schedule tours of their operations for visitors. Tour information is available at the Port of Grays Harbor office in Hoquiam.

In spite of the industrial development along the shores of Grays Harbor, there are still large populations of salmon, sea-run trout, sturgeon, and shad. Tideflats, shoal islands, and wetlands found within the harbor make the area an important habitat for birds and small animals. The Johns River Habitat Management Area along the south shore of the harbor comprises 12,000 acres that protect the Johns River estuary, and is home for black-tailed deer, black bear, and Roosevelt elk, along with many species of birds.

Bowerman Basin, on the north shore of the harbor just west of Hoquiam, is a small tidal mudflat famous for the huge populations of shorebirds that feed and rest there every spring during their migration. (See Shorebirds and Bowerman Basin).

The outer Pacific Coast beach north of Grays Harbor can be reached by heading northwest out of Aberdeen. This section of beach includes the Ocean Shores resort and residential development, which occupies most of the peninsula at the entrance of Grays Harbor, and the largely isolated stretch of beach that reaches north to the Quinault Indian Reservation. This wide gray sandy beach is backed by sand dunes along its southern portions, and broken up by rock outcroppings and sea stacks to the north. Keep an eye out for bald eagles soaring overhead.

The stretch of beach on the outer coast southwest of Aberdeen is also wide and sandy with particularly outstanding examples of primary sand dunes. The beach reaches north to the town of Westport, which sits on the southern peninsula at the entrance to Grays Harbor. Westport, a popular fishing town, boasts one of the best equipped charter boat fleets in the world. Specific information on fishing charters is available by calling the Westport Chamber of Commerce.

At both the north and south entrance to Grays Harbor there are long rock jetties reaching out into the ocean. The jetties are excellent spots to look for seals and sea lions playing in the surf, go sport fishing, or just to take a walk on the rocks.

Driving is permitted on the beaches of Grays Harbor County, but it is not recommended. Automobile traffic damages species that live under the sand, such as the razor clam, and disturbs birds, particularly shorebirds, that live and feed along the beach. If you must drive on the beach drive a safe distance away from the lower beach, which is where razor clams live, and stay out of any dune development along the upper beach.

GRAYS HARBOR COUNTY
NORTH

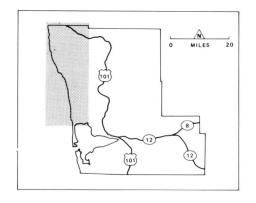
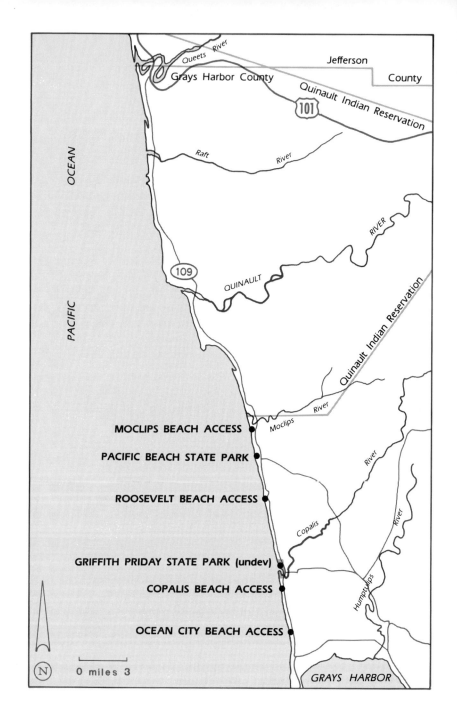

GRAYS HARBOR COUNTY
NORTH

	ACRES	CAMP UNITS	PICNIC UNITS	RESTROOMS	FIREPITS	SWIMMING BEACH	BOAT LAUNCH (lanes)	BOAT MOORAGE (slips/buoys)	PUBLIC PIER	DRINKING WATER	VIEWPOINT	SHORELINE LENGTH (feet)	ROCK BEACH	SAND BEACH	GRAVEL BEACH	MUD BEACH	SAND DUNES	TIDE POOLS	WETLANDS	BLUFFS
COPALIS BEACH ACCESS	NA		•									NA	•		•					
GRIFFITH PRIDAY STATE PARK (undev)	358.7											7,776		•						
MOCLIPS BEACH ACCESS	NA											NA		•						
OCEAN CITY BEACH ACCESS	7.5		•									400		•		•				
PACIFIC BEACH STATE PARK	9.0	40	4	•								1,320		•						
ROOSEVELT BEACH ACCESS	NA		1	•							•	NA		•						

COPALIS BEACH ACCESS
Take Highway 115 north to Copalis Beach; go west on Heath Road to access.

GRIFFITH PRIDAY STATE PARK (undev)
Located on Brenner Road at the north end of the town of Copalis. Undeveloped park with no facilities other than a beach access road.

MOCLIPS BEACH ACCESS
Located at Moclips. Take 2nd Ave. north of town and follow signs. No facilities.

OCEAN CITY BEACH ACCESS
Follow signs in Ocean City. Go west on 1st Street to beach.

PACIFIC BEACH STATE PARK
Located in the town of Pacific Beach. Vehicle access to the beach. Camping in the lot next to the beach.

ROOSEVELT BEACH ACCESS
Located 3 miles south of Pacific Beach; take Highway 109 to signs. Vehicle access to the beach.

GRAYS HARBOR COUNTY
SOUTH

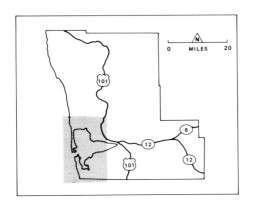
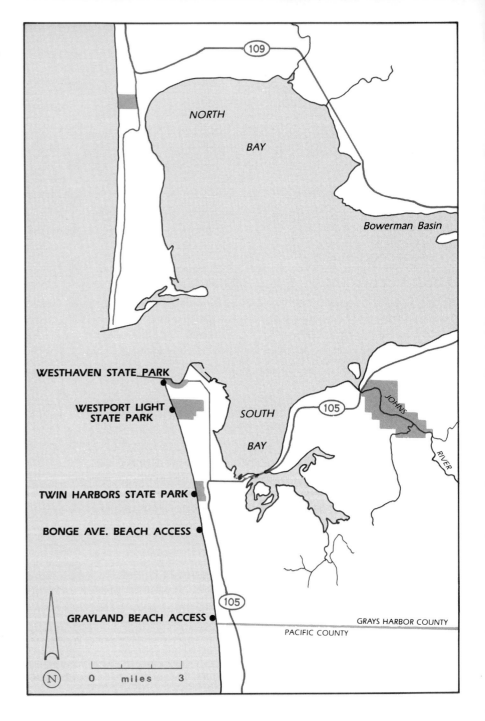

GRAYS HARBOR COUNTY
SOUTH

	ACRES	CAMP UNITS	PICNIC UNITS	RESTROOMS	FIREPITS	SWIMMING BEACH	BOAT LAUNCH (lanes)	BOAT MOORAGE (slips/buoys)	PUBLIC PIER	DRINKING WATER	VIEWPOINT	SHORELINE LENGTH (feet)	ROCK BEACH	SAND BEACH	GRAVEL BEACH	MUD BEACH	SAND DUNES	TIDE POOLS	WETLANDS	BLUFFS
BONGE AVE. BEACH ACCESS	NA		1		•							NA		•			•			
GRAYLAND BEACH ACCESS	NA		•									NA		•			•			
TWIN HARBORS STATE PARK	168.3	338	73	•	•							3,414		•			•			
WESTHAVEN STATE PARK	79.1		12	•	•						•	1,215		•						
WESTPORT LIGHT STATE PARK	212.2		15	•		•						3,397		•			•		•	

BONGE AVE. BEACH ACCESS
Follow Highway 105 0.5 mile south of Twin Harbors State Park; go west on Bonge Road. No facilities. Access point for vehicles.

GRAYLAND BEACH ACCESS
Located 4 miles south of Twin Harbors State Park; go west on Grayland Beach Road. Beach access road only.

TWIN HARBORS STATE PARK
Located 20 miles southwest of Aberdeen on Highway 105. Protected picnic areas near the dunes and beach. Large campground with some R.V. hookup units across the highway from the beach. Another camping area is located on the beach side of the road (no hookup's).

WESTHAVEN STATE PARK
Located just south of Westport off the main road to Westport. There is a sign at the entrance road. Access to south jetty. No vehicle access to beach.

WESTPORT LIGHT STATE PARK
Go north on Highway 105 towards Westport; turn west on West Ocean Ave. to the park and a beach access. Windbreaks around the picnic tables.

GRAYS HARBOR COUNTY
SOUTH

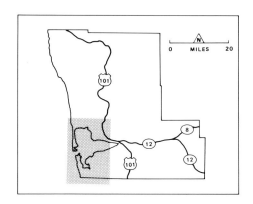
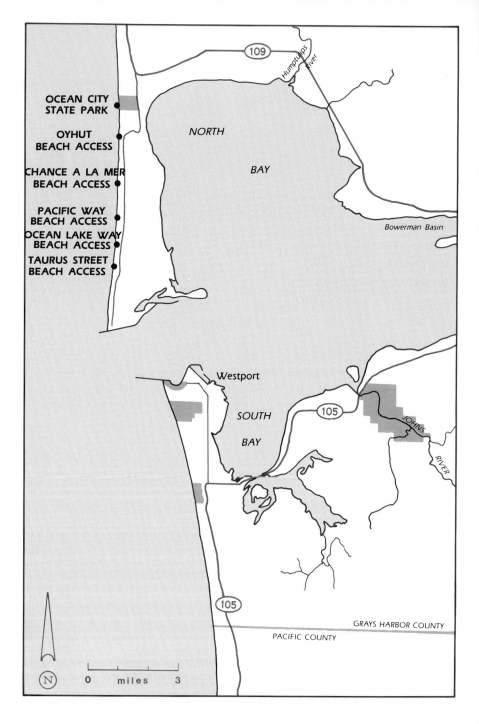

GRAYS HARBOR COUNTY
SOUTH

	ACRES	CAMP UNITS	PICNIC UNITS	RESTROOMS	FIREPITS	SWIMMING BEACH	BOAT LAUNCH (lanes)	BOAT MOORAGE (slips/buoys)	PUBLIC PIER	DRINKING WATER	VIEWPOINT	SHORELINE LENGTH (feet)	ROCK BEACH	SAND BEACH	GRAVEL BEACH	MUD BEACH	SAND DUNES	TIDE POOLS	WETLANDS	BLUFFS
CHANCE A LA MER BEACH ACCESS	NA		•									NA	•		•					
OCEAN CITY STATE PARK	111.8	187	6	•	•	•						2,100	•		•		•			
OCEAN LAKE WAY BEACH ACCESS	NA											NA	•		•					
OYHUT BEACH ACCESS	NA		•									NA	•		•					
PACIFIC WAY BEACH ACCESS	NA											NA	•		•					
TAURUS STREET BEACH ACCESS	NA					•						NA	•		•					

CHANCE A LA MER BEACH ACCESS
Located just south of Ocean City off of Ocean Shore Way on Chance a la Mer Street.

OCEAN CITY STATE PARK
Located 2 miles north of Ocean Shores on Highway 115. Forested area with large camping area. There is a trail to wetlands and ponds. Light, portable boats, such as canoes, kayaks and rafts, can be carried along the trail for launch into the ponds.

OCEAN LAKE WAY BEACH ACCESS
Located 1.5 miles south of Ocean Shores; go right on Ocean Lake Way. Beach access only with no facilities.

OYHUT BEACH ACCESS
Located 0.5 mile north of Ocean Shores. Follow signs off Highway 115.

PACIFIC WAY BEACH ACCESS
Located on Pacific Way, 1 mile south of Ocean Shores City. No facilities, but access is provided to the beach.

TAURUS STREET BEACH ACCESS
Located 2.5 miles south of Ocean Shores; go west on Taurus Street. Beach access with no facilities.

GRAYS HARBOR COUNTY
SOUTH

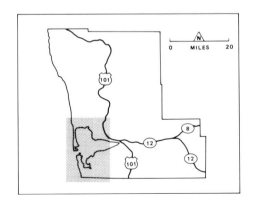
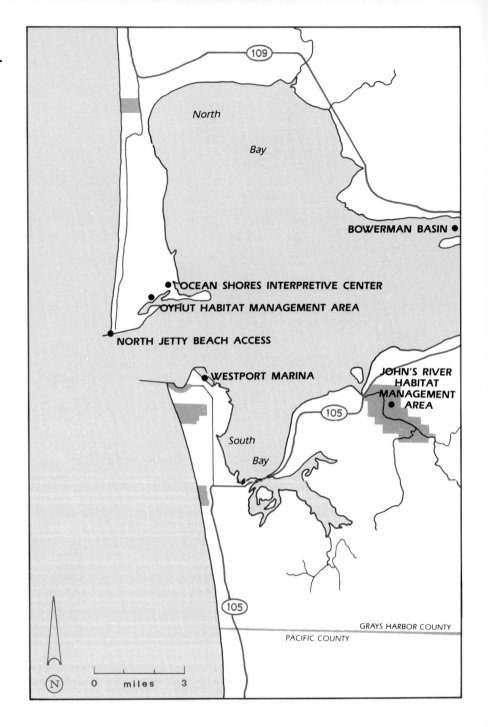

GRAYS HARBOR COUNTY
SOUTH

	ACRES	CAMP UNITS	PICNIC UNITS	RESTROOMS	FIREPITS	SWIMMING BEACH	BOAT LAUNCH (lanes)	BOAT MOORAGE (slips/buoys)	PUBLIC PIER	DRINKING WATER	VIEWPOINT	SHORELINE LENGTH (feet)	ROCK BEACH	SAND BEACH	GRAVEL BEACH	MUD BEACH	SAND DUNES	TIDE POOLS	WETLANDS	BLUFFS
BOWERMAN BASIN	500.0										•	8,000	•		•				•	
JOHN'S RIVER HABITAT MANAGEMENT AREA	15,284.0		•		1							NA			•		•			
NORTH JETTY BEACH ACCESS	NA		•								•	NA	•		•					
OCEAN SHORES INTERPRETIVE CENTER	NA											NA								
OYHUT HABITAT MANAGEMENT AREA	682.5											NA	•		•		•			
WESTPORT MARINA	2.0	4	•		•		3		•			60								

BOWERMAN BASIN
Located west of the city limits of Hoquiam. Take the road marked Bowerman Airport. One of the most significant stopover areas for migratory shorebirds. Best time for seeing birds is mid-April to mid-May. See write-up on Bowerman Basin.

JOHN'S RIVER HABITAT MANAGEMENT AREA
Located 12 miles southwest of Aberdeen on the Westport Highway. Popular as a waterfowl and upland bird hunting area. Large estuarine system includes the John's River from its mouth to 4 miles upstream. A wide variety of coastal birds and animals live here including blacktail deer, Roosevelt elk and black bear.

NORTH JETTY BEACH ACCESS
Located at the south end of Ocean Shores Blvd. The rock jetty provides pedestrian access. Many birds, harbor seals and sea lions.

OCEAN SHORES INTERPRETIVE CENTER
Interpretive center located southeast of Ocean Shores on Grays Harbor. The site of the wrecked steamship the "Catala" was located just south of here on Damon Point.

OYHUT HABITAT MANAGEMENT AREA
Waterfowl habitat management area located at the southeast end of the Ocean Shores peninsula. Features surf fishing as well as hunting for waterfowl. This is one of the first resting stops for Dusty Canada geese on their southerly, fall migration. A wide variety of shorebirds are also found here.

WESTPORT MARINA
Take Montesano Ave. out of Westport and follow to the end. Fishing boardwalk, observation tower with platform and boat launch. Passenger ferry across the harbor to Ocean Shores. There are many fishing boat charter businesses operating out of the marina. The marina has a boat launch.

SHOREBIRDS

The small camouflaged birds with long legs seen scurrying along the water's edge are collectively known as shorebirds. About eighty species of shorebirds live or visit on the west coast. Among the most common in western Washington are sandpipers, dunlin, dowichers, sanderlings, phalaropees, and some small groups of plovers and oystercatchers. The birds are usually seen harvesting Washington's beaches during their migrations between South America and Alaska in the spring and late summer.

Shorebirds are most visible in the air. They rise in unison in huge dense flocks and travel through the air in tight wavy patterns. As they twist and turn in unexplainable harmony, light flashes off their white underbellies, contrasting with their dark backs. Following this incredible display, they slowly settle back onto the beach, all facing the same direction as if waiting for a conductor to orchestrate their next move.

On the beach, the small birds are not as easily spotted. Their mottled brown backs blend in with the sand and mudflats, while their white undersides pick up the color of the ground. What appears to be a barren beach or mudflat may actually be teeming with shorebirds who are enjoying a gourmet feast in a protected haven.

Although the shape and size of shorebirds varies among species, they typically have long legs for wading and long slender bills for probing and searching the shoreline for food. Different birds have specialized physical characteristics adapted for feeding and living on different parts of the beach: The longbilled dowitcher uses its thin long bill to probe in shallow water and deep into the mud for worms, while the black-bellied plover prefers to use its shorter, stronger bill to crack and eat small mollusks and insects on the surface of the beach. These different adaptations allow many species of shorebirds to feed on the same beach without having to compete for food.

Shorebirds do, however, face competition with man. Although they are protected against hunters under the Migratory Bird Treaty Act of 1819, their populations are facing a serious threat due to habitat destruction from shoreline development. Wetlands are being filled, beaches are being altered and polluted, and some shores are crowded with human visitors. The shorebirds are finding fewer natural areas suitable for nesting, feeding, and resting. Visitors to the beach can be sensitive to this problem by keeping their distance from shorebirds.

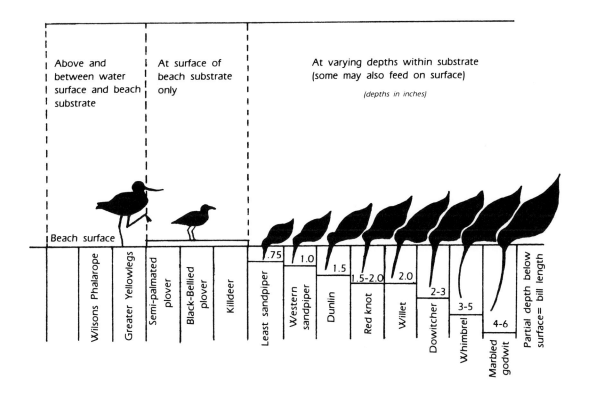

Feeding distribution on mudflats and nearshore waters by selected shorebirds

BOWERMAN BASIN

From mid-April to mid-May, Washington's ocean beaches and tidal mudflats are harvested by myriads of migrating shorebirds. Many of these birds, some from as far south as Chile, arrive at Bowerman Basin to feast in the ninety-four-square-mile Grays Harbor estuary. The estuary is the last major "staging area" before the birds begin their long haul to nesting grounds in the arctic north. Because arctic summers are so short, migration must take place during a brief period of time. For the past three years, approximately 50 percent of the total population of shorebirds visiting Washington has arrived in Grays Harbor on April 24 or 25.

Bowerman Basin comprises only 2 percent of Grays Harbor's area. It is host to such a large percentage of birds because it is an expansive shallow mudflat that provides maximum feeding opportunity: it is the last mudflat to be covered when the tide comes in, and the first to be exposed when the tide recedes.

Visitors to Bowerman Basin during this time will see many species of birds, but the most common are western sandpipers, dunlins, dowitchers, and red knots. It's a good idea to arm yourself with binoculars and a bird guide when visiting Bowerman Basin.

PACIFIC COUNTY

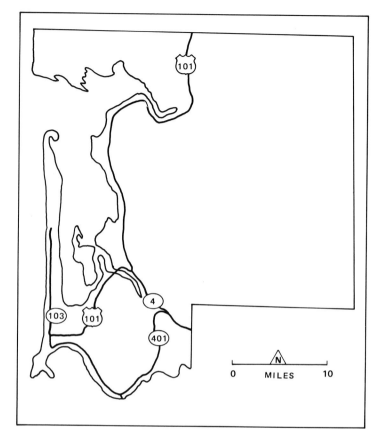

The majority of Pacific County's 155 miles of shoreline surrounds Willapa Bay, which, in addition to having some of the state's most spectacular scenery, is one of the most important estuaries on the West Coast. An ideal oyster growing habitat, Willapa Bay is the largest oyster producing area on the West Coast. In addition it is used by thousands of ducks and geese as a wintering, resting, and feeding area. Long Island, in the middle of the bay, is home for deer, bear, elk, and occasionally bobcats and coyotes.

Because of the array of wildlife in this area, twenty thousand acres of southern Willapa Bay have been set aside as the Willapa National Wildlife Refuge. The refuge has several access points around the bay and is an excellent place to watch birds or just enjoy the natural environment.

The mud and gravel beaches and saltgrass shores of Willapa Bay are broken up by small quiet towns, wetlands, and tidal flats. Many of the towns, which survive mainly on logging and oyster operations, have historical points of interest for visitors.

In the southwest corner of the county, Ilwaco is home port for Columbia River and Pacific Ocean fishing fleets, with moorage space for over a thousand boats. Charters for tuna, salmon, and other deep-sea fishing, and Columbia River sturgeon fishing are available. Some operators also offer scenic and wildlife tours of the Columbia River and Pacific Ocean.

The southern part of the county has two large forts along the shore of the Columbia, Fort Canby and Fort Columbia, which were built to guard the entrance to the river. Both forts have interpretive centers that are open to visitors during the summer. There are also several smaller historic points of interest along the shore.

North of Ilwaco is the Long Beach Peninsula, a huge spit formed by the sediments delivered to the ocean by the Columbia River (See Spits). Once delivered to the ocean, waves and currents slowly reworked the sediments to form the nineteen-mile-long peninsula which is only about a mile wide. The long arm stretching north now protects Willapa Bay from the rough ocean surf. Beautiful sandy beaches make the Long Beach Peninsula one of the most popular recreation areas in the state.

Leadbetter Point State Park and Natural Area is at the northernmost and most recently formed part of the peninsula. The park, which abuts the National Wildlife Refuge, has rich habitats of grasslands, ponds, marshes, forests, small shrubs, and sand. It is an important resting and feeding area for the black brant, a sea goose that arrives by the thousands in April and May enroute to Mexico from Alaska. The outer beach shelters the fragile nesting area of the snowy plover, now a protected species.

Access to the outer ocean beach is available along the entire length of the peninsula. Driving is allowed along the length of the beach, but not recommended because of the detrimental effect on the small marine organisims, clams, and birds that live there.

DUNES

Sand dunes are unique and fragile coastal environments that create challenging conditions for a variety of hardy plants and animals to live in. Species adapted to living on dunes must be tolerant of wind, sand burial, sand abrasion, salt spray, water deprivation, and salty shifting soils. Unfortunately, dunes often conflict with beach recreation and human development; consequently, they are slowly being obliterated. A close look at dunes proves them not to be only fascinating, but also important components of the natural ecosystem.

Most sand dunes in Washington can be found in Grays Harbor County, along the Long Beach Peninsula in Pacific County, at Cattle Point on San Juan Island in San Juan County, and at Deception Pass on Whidbey Island in Island County. The Long Beach dunes, by far the largest dune fields, cover nineteen miles along the peninsula.

Because sand dunes go through a distinct series of developmental changes over time, they offer one of the clearest pictures of successional growth in the natural community. Habitat and species increase in both density and diversity from the youngest dunes, which are closest to the water, to the oldest, which are most landward.

Walking across a dune from the ocean toward the land enables you to encounter three major successional zones of the dune system:

1) The foredune is immediately above the high tide line and has conditions which are severe and unfavorable for plant growth. Typical species that are adapted to this environment are dune wildrye, seashore bluegrass, European beachgrass, and sea rocket. As these varieties become established, loose sand is begining to be stabilized by their extensive root systems, and blowing sand is deposited around the base of the plants, forming low mounds. Beyond these embryo dunes, plant density is able to increase, resulting in greater sand deposition and stabilization. As the environmental conditions become more moderate, several additional species become established.

2) Beyond the foredune is the unstabilized dune. Vegetative cover here is increased but is still not enough to completely stabilize the sand. The dunes in this zone are constantly shifting and subjecting plants to burial or uprooting. Most of the previously mentioned species occur here, along with beach strawberry, bighead sedge, sea purslane, beach morning-glory, sand verbena, and sandmat. Species that live on top of and along the sides

Dune Plant Habitat

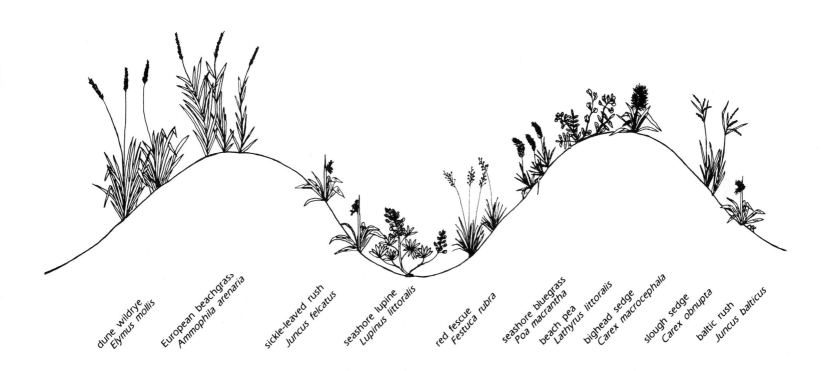

of these dunes get lots of exposure and are different than species living between the dunes in the dune hollow. Often the low spots of the dune hollows are near the water table and hold standing water for several months of the year. As a result, wetland species are often characteristic of dune hollows, which form a microhabitat that is in striking contrast to the dry and exposed species on top of the dune.

3) The stabilized dune zone has nearly solid vegetative cover. In addition to the previously mentioned species, seaside tansy, seashore lupine, red fescue, licorice fern, black knotweed, and sea thrift are common on the dunes. The dune hollows here support dense stands of slough sedge, rushes, and Pacific silverweed. Furthur landward, more protection and stabilization is offered and upland forest vegetation begins to blend in.

The dune habitat, particularly along Long Beach Peninsula, is one of the most outstanding bird watching areas in Washington. An unusual bird that breeds on the outer coast dunes is the snowy plover. The nest sites (which are off limits to people) at Leadbetter Point are the northernmost breeding sites for these birds along the Pacific Coast and the only known sites in Washington.

Other unusual breeding birds may include ring necked ducks, northern shovelers, blue-winged teal, and the cinnamon teal, which nests in dune hollows. Studies have shown that these areas are productive breeding sites for other waterfowl as well.

Recreational use of beach dunes is very heavy in many parts of the state, especially on the Long Beach Peninsula. Dunes seem to tolerate foot traffic fairly well, but trails should be used to avoid trampling the vegetation. The greater recreational impact on dunes comes from off-road-vehicle use. Driving over dunes will destroy the fragile vegetation and allow sand to be lost to wind and storms. Eventually, erosion of the entire dune may cause it to revert back to the earliest stages of succession. It may take years, if ever, for vehicle tire damaged areas to recover.

Dune ridges along the beach can provide valuable protection from storms to low-lying inland areas. The dune ridge provides a buffer of sand that absorbs wave and wind energy. Loss of these ridges not only exposes the natural environment to harsh winds and salt spray, but also threatens human development by letting winter storm waves blow sand and salt water to buildings and roadways.

Dunes are an important part of the beach system. Like almost all features in nature they have a special purpose. Rather than tamper with the carefully balanced dune environment, we should appreciate and enjoy it for its unique ecology.

PACIFIC COUNTY
NORTH

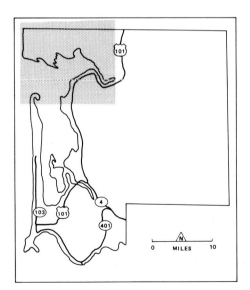

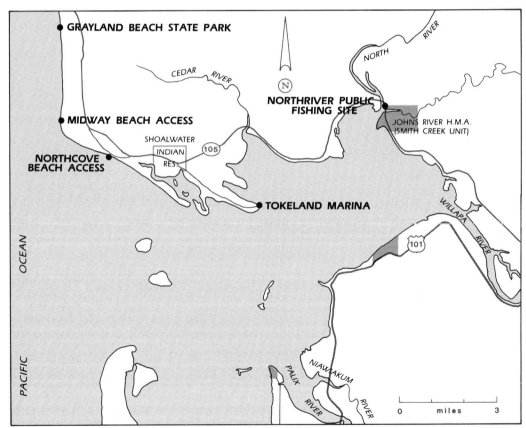

PACIFIC COUNTY
NORTH

	ACRES	CAMP UNITS	PICNIC UNITS	RESTROOMS	FIREPITS	SWIMMING BEACH	BOAT LAUNCH (lanes)	BOAT MOORAGE (slips/buoys)	PUBLIC PIER	DRINKING WATER	VIEW/POINT	SHORELINE LENGTH (feet)	ROCK BEACH	SAND BEACH	GRAVEL BEACH	MUD BEACH	SAND DUNES	TIDE POOLS	WETLANDS	BLUFFS
GRAYLAND BEACH STATE PARK	401.8	62	•	•	•					•		7,509		•			•			
MIDWAY BEACH ACCESS	NA											NA		•						
NORTHCOVE BEACH ACCESS	NA											NA		•						
NORTHRIVER PUBLIC FISHING SITE	NA		•				1			•		100			•					
TOKELAND MARINA	10.0		•				1	20	•	•	•	600		•						

GRAYLAND BEACH STATE PARK
Located 5 miles south of Twin Harbors on Highway 105; 1 mile south of the town of Grayland go right on County Line Road and follow signs. Vehicle access to the beach is at the end of County Line Road. Campsites are on the east side of dunes. Pacific razor clam digging.

MIDWAY BEACH ACCESS
Located 4.5 miles south of Grayland on Highway 105 at Midway Beach. The access point has a sign.

NORTHCOVE BEACH ACCESS
Located 5.5 miles south of Grayland off of Highway 105. Turn on Warrenton Cannery road and go 1 mile to the beach.

NORTH RIVER PUBLIC FISHING SITE
A boat launch into the North River which provides access to upper Willapa Bay. No overnight parking or camping. A State Game Department Conservation License is required to use this facility.

TOKELAND MARINA
Located at Tokeland on Willapa Harbor. Provides a boat launch, and scenic view, but no overnight parking. There is a private overnight park nearby.

PACIFIC COUNTY
NORTH

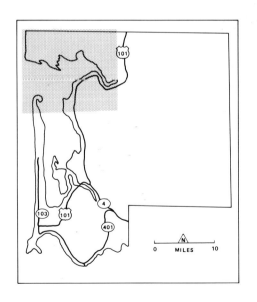

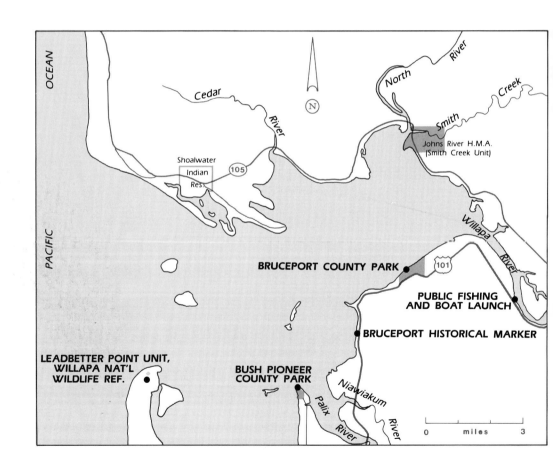

PACIFIC COUNTY
NORTH

PUBLIC SHORE

	ACRES	CAMP UNITS	PICNIC UNITS	RESTROOMS	FIREPITS	SWIMMING BEACH	BOAT LAUNCH (lanes)	BOAT MOORAGE (slips/buoys)	PUBLIC PIER	DRINKING WATER	VIEWPOINT	SHORELINE LENGTH (feet)	ROCK BEACH	SAND BEACH	GRAVEL BEACH	MUD BEACH	SAND DUNES	TIDE POOLS	WETLANDS	BLUFFS
BRUCEPORT COUNTY PARK	57.7	50	15	●	●					●		3,740		●	●				●	●
BRUCEPORT HISTORICAL MARKER	NA									●		NA								●
BUSH PIONEER COUNTY PARK	45.6	50	15	●	●							4,553	●		●				●	●
LEADBETTER POINT UNIT, WILLAPA NAT'L WILDLIFE REF.	1,500.0		●									NA		●			●		●	
PUBLIC FISHING AND BOAT LAUNCH	NA						1					100				●				

BRUCEPORT COUNTY PARK
Located 6 miles west of South Bend on Highway 101. Camping on bluff overlooking Willapa Bay. No beach access.

BRUCEPORT HISTORICAL MARKER
Located 10 miles south of South Bend on Highway 101. Historical marker near the site of historic Bruceville-Bruceport, ca 1851-1880. This area was a center for oystering.

BUSH PIONEER COUNTY PARK
Take Bay Center Road off Highway 101 and follow to the end. The park has gravel bluffs.

LEADBETTER POINT UNIT, WILLAPA NAT'L WILDLIFE REF.
Located at the end of Stackpole Road on the northern tip of the Long Beach Peninsula. Recreational activities include wildlife observation and photography, hiking, surf fishing and waterfowl hunting. Important Salicornia marsh. A portion of the area is closed to all public access from April 1 through August 31 to protect nesting snowy plovers.

PUBLIC FISHING AND BOAT LAUNCH
Located 1 mile west of South Bend on right side of Highway 101. No facilities. A Department of Game Conservation License is required to use this launch.

PACIFIC COUNTY
CENTER

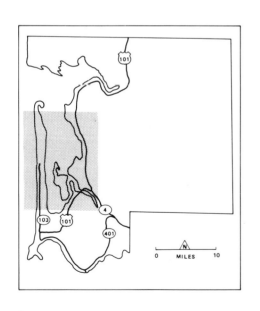
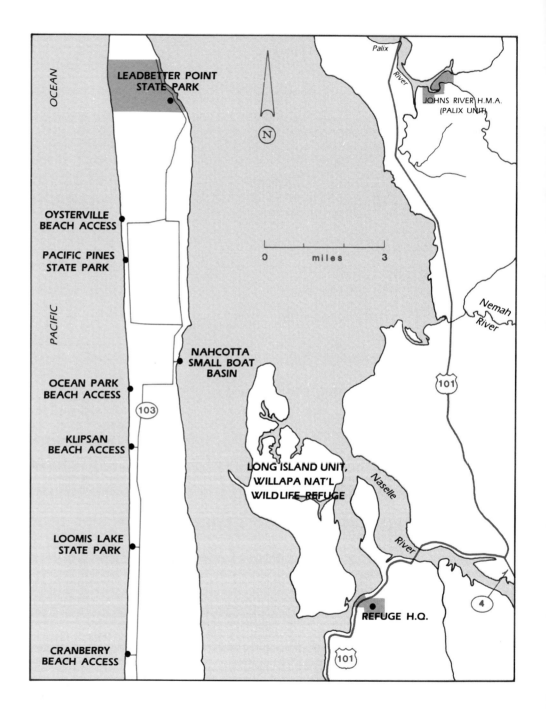

PACIFIC COUNTY
CENTER

PUBLIC SHORE

	ACRES	CAMP UNITS	PICNIC UNITS	RESTROOMS	FIREPITS	SWIMMING BEACH	BOAT LAUNCH (lanes)	BOAT MOORAGE (slips/buoys)	PUBLIC PIER	DRINKING WATER	VIEWPOINT	SHORELINE LENGTH (feet)	ROCK BEACH	SAND BEACH	GRAVEL BEACH	MUD BEACH	SAND DUNES	TIDE POOLS	WETLANDS	BLUFFS
CRANBERRY BEACH ACCESS	1.6		•									225		•			•			
KLIPSAN BEACH ACCESS	NA		•									NA		•			•			
LEADBETTER POINT STATE PARK	807.3		•								•	15,840		•			•		•	
LONG ISLAND UNIT, WILLAPA NAT'L WILDLIFE REFUGE	23,300.0	20	•				1					132,000	•	•		•	•			
LOOMIS LAKE STATE PARK	13.5		24	•	•							425		•			•			
NAHCOTTA SMALL BOAT BASIN	NA		•					100	•	•		1,000			•					
OCEAN PARK BEACH ACCESS	NA											NA		•			•			
OYSTERVILLE BEACH ACCESS	NA											NA		•			•			
PACIFIC PINES STATE PARK	10.8		15	•	•							590		•			•			

CRANBERRY BEACH ACCESS
Located 3 miles north of Long Beach off of Highway 103.

KLIPSAN BEACH ACCESS
Located 8.5 miles north of the town of Long Beach on the peninsula.

LEADBETTER POINT STATE PARK
Located at the northern tip of the Long Beach Peninsula, 3.0 miles north of Oysterville on Stackpole Road. This state park natural area is a unique blend of areas that include shifting dunes, grasslands, ponds, marshes and forests. An abundance of plant and animal life can be found in these diverse habitats, including approximately 100 species of birds. A portion of the park along the ocean beach side is closed to all entry from April through August to protect the nesting snowy plovers. Camping is prohibited.

LONG ISLAND UNIT, WILLAPA NAT'L WILDLIFE REFUGE
Access to the island is via the boat launch at refuge headquarters 13 miles north of Ilwaco on Highway 101. Five campgrounds located along the shoreline of the island are accessible only by boat. No potable water is available. The island is forested. Permitted recreational activities include wildlife observation, hiking, camping and archery hunting for big game and upland birds. Wildlife includes black bear, Roosevelt Elk, black-tailed deer, blue and ruffed grouse, bald eagles, shorebirds, and waterfowl.

LOOMIS LAKE STATE PARK
Located 5 miles north of the town of Long Beach off of Highway 103. Sheltered picnic units behind sand dunes and trail to the beach.

NAHCOTTA SMALL BOAT BASIN
Located on Sandridge Road on the Long Beach Peninsula. Provides boat launching into Willapa Bay.

OCEAN PARK BEACH ACCESS
Located in the town of Ocean Park on the Long Beach Peninsula.

OYSTERVILLE BEACH ACCESS
Provides access to the ocean directly opposite the town of Oysterville on the Long Beach Peninsula.

PACIFIC PINES STATE PARK
Located 1 mile north of Ocean Park on the Long Beach Peninsula. Provides a picnic area nestled among shore pine trees. No overnight camping.

WILLAPA NATIONAL WILDLIFE REFUGE

The Willapa National Wildlife Refuge, encompasing a huge estuary in Pacific County, was established in 1937 to provide a haven for hundreds species of birds and animals. The estuary's diverse environment includes protected bays, islands, wetlands, eelgrass beds, and mudflats. Nutrients carried by the incoming tides mix with nutrients from rivers to make the estuary highly productive. Some of the more visible animals that take advantage of the refuge are seals, waterfowl, eagles, bear, deer, and elk.

The best time to see wildlife at Willapa is during the winter when birds travel south from Alaska to the protected and nutritious waters of the bay. Keep an eye out for large flocks of black brant, Canada geese, American widgeons, canvasbacks, scaups, buffleheads, and scoters. Loons, grebes, mergansers, and cormorants are also out in the bay, and dunlins, plovers, and sandpipers line the shore. During fall and spring migrations even larger concentrations of waterfowl and shorebirds pass through the refuge.

Birds that nest in the refuge during spring and summer include grouse, bald eagles, herons, woodpeckers, and shorebirds. A portion of Leadbetter Point is closed to the public in the summer to protect the nesting grounds of the snowy plover, a shorebird particularly sensitive to human disturbance.

There are four access areas to Willapa Refuge:

1) Leadbetter Point access is located at the northern tip of the Long Beach Peninsula. The point is an area of transitory sand dunes, wetlands, and mudflats that is a particularly popular spot for shorebirds. Bird watching is best here during migration.

2) Long Island, located in the southeast corner of the bay, is a densely forested and often rain-drenched island which is accessible only by boat. It has several campgrounds. Deer, elk, grouse, beaver, and numerous small mammals and songbirds make their home here. A stand of huge old-growth western red cedar is a remnant of the mature forest that once covered the island. When crossing the bay to Long Island, be aware of outgoing tides that could leave you trapped on the mudflats.

3) The Lewis Unit access area off Highway 101 takes visitors to the freshwater marsh at the south end of the bay. A small flock of trumpeter swans spends every winter here.

4) Riekkola Unit access is also at the south end of the bay. Here extensive grasslands on diked tidelands provide popular feeding areas for migrating Canada geese, ducks, and shorebirds.

Popular activities at Willapa Bay are hiking, boating, clamming, crabbing, and fishing. The State Shellfish Lab at Nahcotta has information and maps on shellfish gathering Certain areas of the refuge are open to hunting waterfowl, upland game birds, and big game. Consult the refuge manager for current hunting regulations. Camping and fires are permitted only in designated campgrounds. When going into the refuge, carry food and water. Binoculars and a bird guide are also a must for bird watchers.

For more information contact:

Refuge Manager
Willapa National Wildlife Refuge
Ilwaco, WA 98624
(206)484-3482

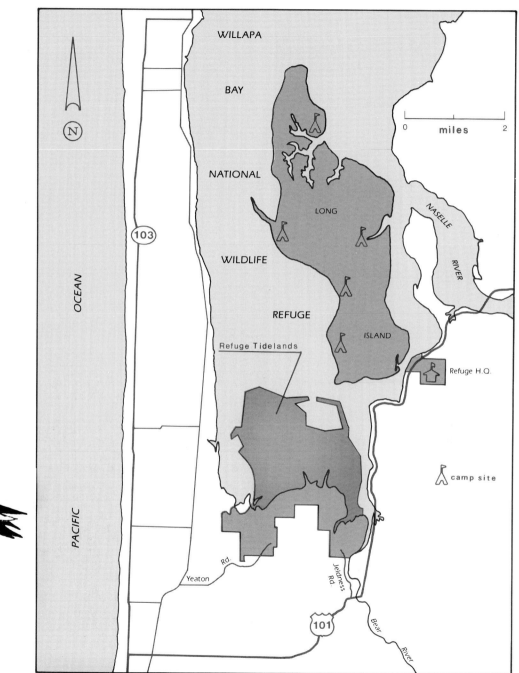

PACIFIC COUNTY
SOUTH

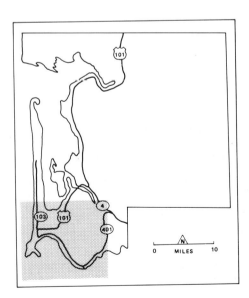

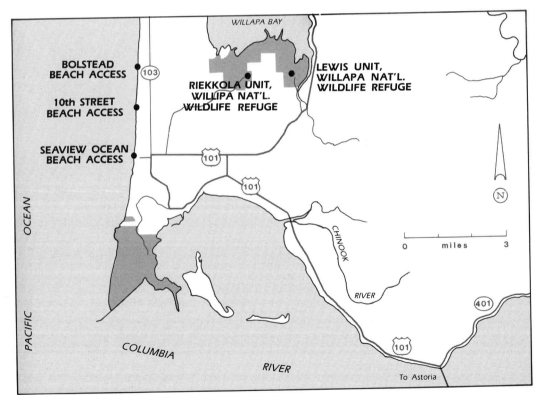

PACIFIC COUNTY
SOUTH

	ACRES	CAMP UNITS	PICNIC UNITS	RESTROOMS	FIREPITS	SWIMMING BEACH	BOAT LAUNCH (lanes)	BOAT MOORAGE (slips/buoys)	PUBLIC PIER	DRINKING WATER	VIEWPOINT	SHORELINE LENGTH (feet)	ROCK BEACH	SAND BEACH	GRAVEL BEACH	MUD BEACH	SAND DUNES	TIDE POOLS	WETLANDS	BLUFFS
10th STREET BEACH ACCESS	NA											NA	•				•			
BOLSTEAD BEACH ACCESS	NA											NA	•				•			
LEWIS UNIT, WILLAPA NAT'L. WILDLIFE REFUGE	NA		•									NA		•					•	
RIEKKOLA UNIT, WILLAPA NAT'L. WILDLIFE REFUGE	NA		•									NA		•					•	
SEAVIEW OCEAN BEACH ACCESS	NA		•								•	NA	•				•			

10th STREET BEACH ACCESS
Located at the west end of 10th street in Long Beach. Provides a driving access point to the beach.

BOLSTEAD BEACH ACCESS
Located 0.5 mile north of the town of Long Beach. Provides a driving access point to the beach.

LEWIS UNIT, WILLAPA NAT'L. WILDLIFE REFUGE
Located 4 miles east of Long Beach, take Highway 101 east from Seaview to Jeldness Road and go left for less than a mile. This refuge unit has freshwater marshes at the south end of the bay which provide habitat for numerous waterfowl. Foot trails on dikes through the area. Permitted recreational activities include wildlife observation, hiking and waterfowl hunting.

RIEKKOLA UNIT, WILLAPA NAT'L. WILDLIFE REFUGE
Located east of the town of Long Beach near the southern tip of the Long Beach Peninsula; take Sandridge Road to Yeaton Road and go east for 2 miles. The gravel road ends at a field where you can walk to wetlands. Grasslands on diked tidelands provide feeding areas for shorebirds and waterfowl. Permitted recreational activities include wildlife observation and waterfowl hunting.

SEAVIEW OCEAN BEACH ACCESS
Located at the end of 38th Street in the town of Seaview. Provides driving access to Long Beach.

PACIFIC COUNTY
SOUTH

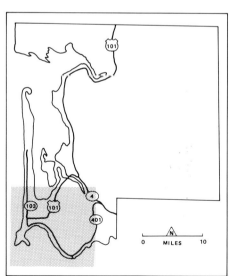
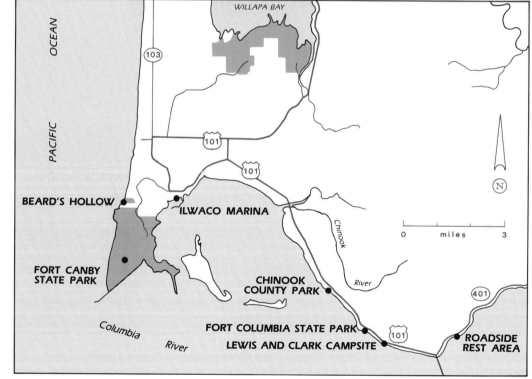

PACIFIC COUNTY
SOUTH

	ACRES	CAMP UNITS	PICNIC UNITS	RESTROOMS	FIREPITS	SWIMMING BEACH	BOAT LAUNCH (lanes)	BOAT MOORAGE (slips/buoys)	PUBLIC PIER	DRINKING WATER	VIEWPOINT	SHORELINE LENGTH (feet)	ROCK BEACH	SAND BEACH	GRAVEL BEACH	MUD BEACH	SAND DUNES	TIDE POOLS	WETLANDS	BLUFFS
BEARD'S HOLLOW	NA											NA		•						•
CHINOOK COUNTY PARK	19.2	50	40	•	•							1,140								
FORT CANBY STATE PARK	1,672.1	254	50	•	•		2	12		•	•	39,000	•		•	•			•	•
FORT COLUMBIA HISTORICAL STATE PARK	591.8			•								6,400								•
ILWACO MARINA	50.0			•				800	•			2,600		•					•	
LEWIS AND CLARK CAMPSITE STATE PARK	0.8	3										NA								
ROADSIDE REST AREA	NA		4	•								NA								

BEARD'S HOLLOW
An undeveloped recent acquisition to Fort Canby State Park which provides opportunities for surf fishing and beach walking. No developed access to this area nor is it possible to get to without walking a long way.

CHINOOK COUNTY PARK
Located off of Highway 101 just east of the town of Chinook. The site is on the Columbia River. An old park with minimal facilities.

FORT CANBY STATE PARK
Fort Canby State Park is located 2 miles southwest of Ilwaco off Highway 101, where the Columbia River meets the Pacific Ocean. This 1700 acre park features four trails of varying lengths which wind through the forests and headlands. Migrating birds are commonly seen along the beach. Clam digging is allowed. A boat launch on Baker Bay has ample parking and a small dock. The Long Beach Peninsula stretches north from the park. The Lewis and Clark Interpretive Center is open 9:00 until 5:00, seven days a week during the summer and by appointment only during the winter. Includes the Beard's Hollow area which is an undeveloped recent addition to the park.

FORT COLUMBIA HISTORICAL STATE PARK
Located 1.0 mile east of Chinook on Highway 101. Park includes an A.Y.H. youth hostel, interpretive center, art center, museum, interpretive trail and a historical walk. Bluffs cause the beach to be unusable.

ILWACO MARINA
Located at the town of Ilwaco. Limited public facilities, but a great place to walk around and look at the fishing fleet.

LEWIS AND CLARK CAMPSITE STATE PARK
Located 2 miles east of Chinook. There is no waterfront. Site is across the highway from the water. Designated as a historical marker, from this site Lewis and Clark saw the breakers of the Pacific Ocean and knew they had reached their destination.

ROADSIDE REST AREA
Located 1 mile east of the Astoria-Megler Bridge on Highway SR-401. Fronts on the Columbia River and has a bulkheaded bank. No beach; fenced on the water side.

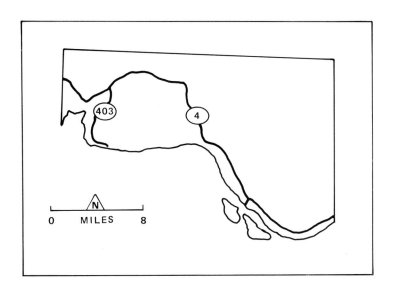

WAHKIAKUM COUNTY

Although Wahkiakum County is seven miles from the ocean, its fifty-five miles of Columbia River shoreline has a strong marine influence. Ocean tides are noticeable along the shore of this small county, the smallest in the state, and large ocean freighters travel back and forth on the Columbia. Like many areas of southwest Washington, the county's economy is based on fishing and lumber.

Steep, forested hills rising out of the river make the area scenic, but actual access to the water difficult. In the eastern half of the county State Highway 4 runs near the shoreline offering good views of Oregon and ocean ships on the Columbia. The road cuts inland at Skomokawa (Ska-mock-away) where the shore is particularly steep and rocky.

The small town of Cathlamet which sits high on the bank of the Columbia, is the county seat. The Wahkiakum County Historical Museum, in addition to many historical buildings are in Cathlamet. About eight miles west of Cathlamet the town of Skamokawa is located along a slough at the mouth of three creeks. Once the third largest town on the river behind Astoria and Portland, Skamokawa is now bordering on a ghost town.

An interesting side trip to take in Wahkiakum County is to travel down Highway 403 to the historic fish cannery town of Altoona. A few people still live here, but the cannery no longer operates.

The Columbia White-tailed Deer National Wildlife Refuge is just out of Skamokawa. The deer is an endangered species found only along the lower Columbia River, and the Umpqua River in Oregon. The deer used to be abundant throughout Washington and Oregon but because of habitat destruction it disappeared from nearly all its range. In the 1930's it was thought to be extinct. Remnant populations were later discovered and in 1972 a sanctuary was established for approximately 230 of the remaining deer. The best time to see the deer is between September and May. Aside from wildlife observation, other activities at the refuge are hiking, boating, hunting, and sport fishing.

WAHKIAKUM COUNTY

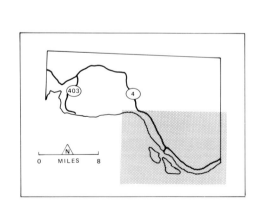

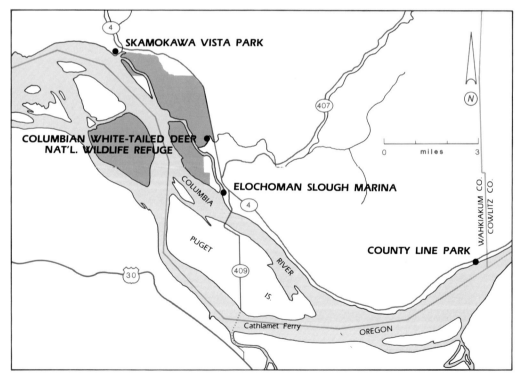

WAHKIAKUM COUNTY

	ACRES	CAMP UNITS	PICNIC UNITS	RESTROOMS	FIREPITS	SWIMMING BEACH	BOAT LAUNCH (lanes)	BOAT MOORAGE (slips/buoys)	PUBLIC PIER	DRINKING WATER	VIEWPOINT	SHORELINE LENGTH (feet)	ROCK BEACH	SAND BEACH	GRAVEL BEACH	MUD BEACH	SAND DUNES	TIDE POOLS	WETLANDS	BLUFFS
COLUMBIAN WHITE-TAILED DEER NAT'L. WILDLIFE REFUGE	4,400.0										•	26,400	•							
COUNTY LINE PARK	6.2		•	•	•					•	•	NA	•							
ELOCHOMAN SLOUGH MARINA	NA			•	•			3		•	•	1,500								
SKAMOKAWA VISTA PARK	35.0	19	13	•	•	•	1			•	•	2,700	•							

COLUMBIAN WHITE-TAILED DEER NAT'L. WILDLIFE REFUGE
From Highway 4 turn off on the Steamboat Slough Road. Wildlife viewing and hiking only. The Columbian white-tailed deer is an endangered species found only at this location and at one location in Oregon. Waterfowl also use this refuge in winter. Sport fishing is permitted from the refuge dike in the Elochoman and Columbia rivers and in the pond at the diking district pump station.

COUNTY LINE PARK
Located next to the county line on the Columbia River 10 miles east of Cathlamet. Beautiful beach and scenic views. Approximately 1200 feet of frontage on the Columbia River. A good place to watch ships go by. Mt. St. Helens is visible.

ELOCHOMAN SLOUGH MARINA
Located in Cathlamet. Provides a protected boat launch and moorage off the Columbia River.

SKAMOKAWA VISTA PARK
Located in Skamokawa off Highway 4 on the Columbia River. Unpatrolled beach.

Drawing by Deanna Hofmann

BEACHES

We typically think of a beach as simply an enjoyable area where the land meets the water. But biologically a beach is a complicated transition zone between terrestrial and marine environments. This transition zone, or ecotone, contains species characteristic of both the environments in addition to some that occur only within the beach zone. As a result, beaches and nearshore areas are highly diversified and productive places, teeming with life.

The types of organisms that inhabit a beach and the complexity of the beach community is controlled by three main factors: (1) the composition of the beach (rock, gravel, sand, mud, or clay); (2) the effect of wind, waves, and currents on the beach; and (3) the amount of time the beach is out of the water at low tide and exposed to sun, heat, drying, freezing, and predation from land animals, including man.

These factors combine in hundreds of different ways to produce an infinite variety of biological life on beaches. For the sake of clarification, we will consider four major types of beaches common along Washington's shoreline: rock, gravel, sand, and mud.

Each type of beach also offers a variety of habitats. If you work your way down a beach from the high-tide line to the low-tide line you will observe that the distribution of the most obvious plants and animals changes with the slope of the beach. Each elevation has a different length of exposure to the tide, which causes a phenomenon known as intertidal zonation. An intertidal zone has characteristic plants and animals that are best adapted to life at that elevation or tidal exposure. Generally, life at the upper zone is determined by the tolerance of the plant or animal to physical effects such as wave energy, desiccation (drying out), or temperature extremes, and life at lower zones are more affected by biological interactions such as predation and competition for space.

As an example, the lower limit of mussel habitat is determined by predation from a starfish called pisaster or purple sea star. Experimental removal of pisaster has shown that mussels are perfectly capable of surviving at lower beach elevations. Pisaster, on the other hand, can only prey on the mussels at the lower part of the mussels' range; moving higher will cause fatal exposure to heat and sun at low tide. In this case the mussels' habitat is defined or limited by the sea star, whose habitat is defined by the tide. Each of these species lives in a different but overlapping intertidal zone.

Each intertidal zone has a community of plants and animals that interact with one another. These communities have overlapping ranges throughout intertidal elevations, although most organisms have their greatest abundance in only one zone. The most conspicuous organisms of a zone are often considered "indicator species" of a particular zone.

The most dependable indicator species of an area are plants and immobile invertebrates because they are usually restricted to a single beach substrate and zone. Fish, birds, and mammals are more adaptable and mobile and although they may prefer a zone on one type of beach, they usually occur in several zones on several beach substrates.

The following is a description of the four major beach types in Washington, their general location, and some of the biological life found in them.

ROCK BEACH

Rock beaches consist of solid bedrock and boulders which are too large to be moved about by currents or wave action. While not composing a major portion of Washington's coastal zone, rock beaches provide some of the most fascinating environments in the state. This is especially true of moderately to highly exposed ocean beaches which are characterized by lush growths of algae and large numbers of animals.

Rock beaches with high or moderate exposure occur along the Pacific Coast of the Olympic Peninsula from Cape Flattery to just south of the Copalis River; along the Strait of Juan de Fuca; on portions of the west side of Whidbey Island; the southwest shore of San Juan Island, and the southern end of Lopez Island. Protected rock beaches are most common in northern Puget Sound. Rock beaches are rare in central and southern Puget Sound, occurring only in a few scattered locations.

Many animals and plants that live on the outer coast rock beaches thrive only in this environment. They develop ways to anchor themselves to the stable rock so they don't get crushed by the waves, and they often form an "addiction" to high levels of oxygen created by pounding waves.

Because waters are quieter in Puget Sound, a stable rock beach environment may consist of smaller boulders than those on the outer coast, where wave action is greater. The

species that live in quieter waters have some different needs and adaptations than those living on the outer coast.

Zonation of Rock Beaches

Intertidal zonation of rock beaches begins at the upper limits with the splash zone. This zone is characterized by a small number of species because the exposure and drying are too extreme for all but the hardiest types. Lichens often form a distinct band here. The rock slater, an isopod which feeds on decaying algae, is most representative of this zone.

The next zone down, called the high intertidal zone, still has relatively few species, though there is more algae here than in the splash zone, the most common being red algae. Two species of filter feeding barnacles also are common. The algae and barnacles together provide habitat for numerous small crustaceans (such as isopods, amphipods, tanaids, and small hermit crabs) and insect larvae.

The mid-intertidal zone is inundated more regularly by tides and therefore sustains more life. On exposed ocean beaches,

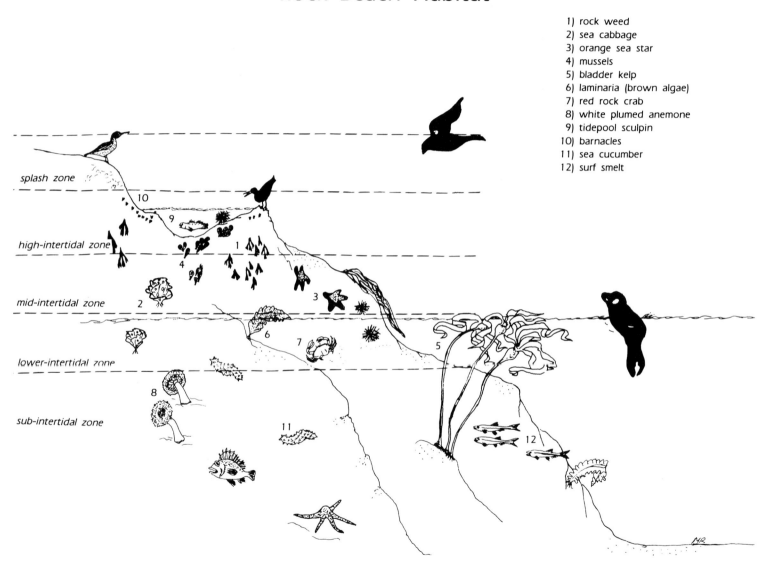

California mussels and gooseneck barnacles invariably dominate this zone. On protected beaches, especially when the slope of the rock is rather steep, there is no characteristic band of mussels due to heavy predation by starfish. In Hood Canal, the Pacific oyster often dominates this zone. Rockweed is usually present in the upper parts of the mid-intertidal zone, especially in protected areas. Larger brown algae, such as the sea cabbage and honey ware kelp, are also present. Sea lettuce may form a distinct dark green band here. Mussels and algae provide a complex miniature "forest" which is literally swarming with small invertebrates, especially crustaceans.

The lower intertidal zone is under water the majority of the time, and therefore offers a less stressful environment to marine life than the higher zones. As a result, it has the highest diversity of plant and animal life and fewer fluctuations in population. Soft bodied animals which cannot tolerate exposure to air are present and the existence of many specialized organisms indicates the complexity of this community. On moderately or highly exposed rock beaches, the biomass of plants in this zone is considerable, due mostly to the presence of kelp. In quieter waters, sea lettuce and other closely related algae may extend well down into the lower intertidal zone.

The dominant animals of the lower intertidal zone are starfish, sea cucumbers, anemones, sea urchins, and certain crabs. The purple sea star is the most typical animal. Sea cucumbers are abundant where there are boulders and crevices and three species of urchins, all herbivores, live here.

Anemones, common crabs, and several types of shrimp also make their home in the lower intertidal zone. Other important herbivores inhabiting the lower intertidal zone include several species of chiton, including the world's largest, plus limpets, snails, barnacles, and occasionally burrowing clams.

Below this zone the shallow subtidal zone begins. The most outstanding feature here is the presence of the larger kelps, which are especially abundant on moderately to highly exposed beaches. Because this area is never uncovered by tides, it is the least stressful.

A variety of mobile animals use rock beaches for nesting, feeding, and resting. Marine mammals such as harbor seals and river otters haul out on these beaches, and some terrestrial mammals such as raccoons, mink, and skunks occasionally forage here. At low tide, garter snakes have been observed feeding on a variety of organisms that live on rock beaches. Many birds, including shorebirds, gulls, diving birds, bald eagles, and peregrine falcons also feed on the abundant life attached to rock shores.

Juvenile salmon feed heavily on small invertebrates living around rock beaches before leaving for their open ocean migration. Herring attach their eggs to intertidal and shallow subtidal vegetation that grows along the shore.

The high diversity of plant and animal life makes rock beach areas particularly desirable for watching wildlife. At the same time, intensive use of these beaches can have detrimental effects on the wildlife. This is particularly noticeable when people collect species for souvenirs, depleting populations and upsetting the ecological balance of the shore. It could take years for a picked-over beach to reestablish a healthy community.

GRAVEL BEACH

The key feature to look for in a gravel beach is wave energy. Where wave action is powerful, the rocks are tumbled and plants and animals are usually crushed. As a result, exposed gravel beaches will almost always be barren. On the other hand, the gravel beaches of Puget Sound's protected waters are generally very productive. A gravel beach can range from a one composed solely of small boulders to a beach with mud, sand, and gravel all mixed together. Gravel beaches are by far the most common beaches in Puget Sound. The following discussion applies to protected gravel beaches.

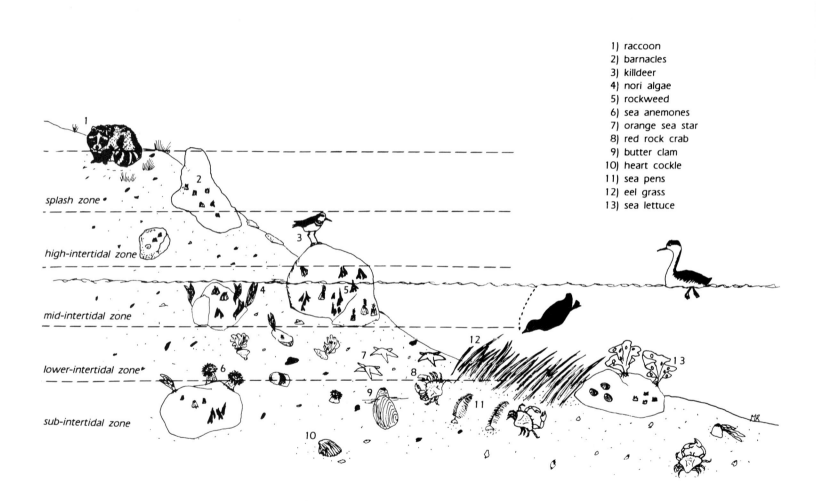

Mixed composition gravel beaches have a high species diversity that results from the variety of substrates present. While boulders and large cobble may support many of the species typical of rock beaches, pockets of sand support animals typical of sand beaches.

In addition, the undersides of rocks and crevices between boulders support several distinct species not typical on other beaches. This type of specialized locale within a larger habitat or community type is called a microhabitat. Microhabitats tend to increase species diversity in an area.

The variety of plants and animals found on mixed gravel beaches will vary widely depending on the specific composition of the beach. A gravel beach with a high percentage of mud and sand will have more vegetation than a coarser gravel beach. Generally, zonation of gravel beaches is not as clear as it is on rock beaches. Species distribution is more affected by substrate and less by tidal exposure.

In the mid-intertidal zone, pockets of sand will provide habitat for a number of organisms, especially burrowing worms and clams. Edible clams which may be found here include butter clams, native littleneck clams, Manila clams, and softshell clams. In Hood Canal and central and southern Puget Sound, bay mussels often occur in large patches, especially in the alluvial fans formed by small creeks flowing down the beach.

Some species that are found on rock beaches occur in greater abundance on mixed gravel beaches. This is because they can find refuge from predators, particularly at low tide, by spending much of their time under rocks. Several species of crabs, such as purple shore crabs, porcelain crabs, and hairy cancer crabs frequently hide under cobble sized rocks.

The six ray, or brooding, starfish is the most common starfish of mixed gravel beaches and green sea urchins reach their highest densities on this beach substrate. Hundreds of isopods and amphipods can often be found hiding out under rocks. Some species of fish also are adapted to intertidal life on gravel beaches and seek protection under rocks while the tide is low. These fish include northern clingfish, penpoint gunnels, high cockscomb, and sometimes sculpins.

On finer composition beaches, the amount of vegetation may increase dramatically in the lower intertidal zone. Eelgrass often grows, along with and the plants and animals associated with it. In well protected areas dense growths of green algae such as sea lettuce are very likely. Several large brown algae such as seersucker, sea collander, and sugar wrack may occur in the lower portions of the zone.

Protected gravel beaches provide excellent foraging and resting habitat for shorebirds and waterfowl, which are especially common in winter. The most common waterfowl seen dotting the water are the common goldeneye, white-winged scoter, red-necked grebe, and European and American widgeon. The killdeer is probably the most abundant shorebird.

Raccoons frequent Puget Sound's gravel beaches to feast on shore crabs and beach hoppers (sand fleas). Even if you don't actually see a raccoon you can probably find tracks in muddy parts of the beach.

Harbor seals will feed close to shore along mixed gravel beaches and have been observed mating in shallow water at more isolated areas. They like to watch human visitors on the beach from a safe distance off shore.

One of the biggest threats to the intertidal mixed gravel beach habitat comes from recreational clam digging. Clam populations are usually monitored by Department of Fisheries officials and when populations get low, "no digging" signs are posted. These signs should be observed at all times. When clam digging is allowed, precautions should be taken, such as filling in holes, to decrease mortality to other clams and vulnerable animals. Not only is this important for clams, but also for the many fish, birds, and crabs which feed on the organisms living there.

MUD BEACH

Mud beaches have been the most misunderstood and abused beach type in Washington. They often seem useless from man's point of view because they are a hindrance to navigation and are not suitable for walking. Biologically, though, mud beaches are very productive. The combination of bacteria and mud can serve as a bountiful food supply for some animals. Depending on the composition of the mud, though, it can also suffocate or immobilize some organisims.

A mud beach substrate consists of fine silt, clay particles, and occasionally some sand. Mud beaches exist only in protected waters because high waves and currents would wash the mud away. A pure mudflat without sand or larger sediments can be very difficult and often dangerous to walk on; one may easily sink well past the knees in soft mud and get stuck.

Because mud beaches are often formed by sediments delivered by rivers and streams, most mudflats receive considerable freshwater influence. Mud beaches typically form broad, almost level tideflats, and have tidal channels created by water draining at low tide. Examples of mud beaches, or mudflats, are found at Mud Bay in Thurston County, Fidalgo Bay in Skagit County, and Bowerman Basin in Grays Harbor County.

Because they occur in calm protected areas, mud beaches are a relatively stable environment. Fine sediments and organic matter hold large amounts of water, mainly by capillary action, and maintain a thin layer of water over the mud at low tide. This moisture helps protect the many small organisms which live within, or at the surface, of the mud.

Mudflats provide an excellent environment for bacteria, and other microorganisms, which are extremely important to the ecology of mudflat communities. Mud typically has a shallow (about one inch) oxygenated layer at the surface, which is a brownish color. Below that is a black non-oxygenated layer of sediments. Digging into the black layer releases small amounts of hydrogen sulfide gas which gives off the odor of rotten eggs, a smell common to mudflats. The gas is produced by anaerobic (without oxygen) bacteria who get chemical energy from sulfur compounds, much in the same way that plants get energy from sunlight. These anaerobic bacteria contribute to the primary production of mudflats, and in turn support large numbers of associated animals.

The density of invertebrates on mud beaches in the high intertidal zone is greater than in any comparable zone on any other beach substrate. This feature is of tremendous importance to vertebrates, especially shorebirds, which feed on the invertebrates. High intertidal areas are exposed for long periods of time at low tide, giving shorebirds ample opportunity to feed. During winter, when the mean tide levels are higher, the birds become even more dependent on the high elevation mudflats.

Waterfowl are also abundant at mud beaches. Green-winged teal, canvasbacks, and ruddy ducks can commonly be seen feeding in these areas. Bird populations and diversity increase significantly during spring and fall migration, which are popular bird watching times around mudflats.

Characteristic large invertebrates found on mudflats include red rock crabs, Dungeness crabs, hairy shore crabs, mud shrimp, Macoma clams, softshell clams, heart cockles, and occasionally Pacific oysters.

Large numbers and varieties of fish feed along mud beaches, especially if eelgrass is present in lower intertidal and shallow subtidal areas. Species include juvenile English sole and starry flounder, three-spine stickleback, shiner perch, sharpnose sculpin, surf smelt, and Pacific herring. In addition, tidal channels provide important feeding areas for salmon.

The impact of toxins in the water is particularly felt on mud beaches. Because these areas are normally very protected and have restricted movement of water, toxins are not as diluted and tend to concentrate more quickly than in other areas. In some parts of southern Puget Sound, water takes more than twenty years to completely recycle. Pollutants discharged into the water in small but constant quantities can accumulate to extremely high levels over long periods of time. The buildup of coliform bacteria can make it unsafe for humans to consume shellfish, particularly oysters, and can jeopardize shellfish industries which depend on these areas.

SAND BEACH

Although most people find surf-pounded sandy beaches to be the most attractive kind, most other forms of life find sand beaches the least attractive. Only a handful of exceptionally hardy species can withstand the perpetual pounding of waves on the unstable sand. Because the main protection found on a sand beach is under the sand, most species that live here are efficient burrowers. Sandy beaches in protected areas usually host a greater number of inhabitants than exposed beaches.

Exposed sand beaches are associated with high waves and wind and strong current. Many beaches along the outer Pacific coast, especially south of Point Grenville are in this category. On relatively calm days the waves on these beaches are

usually three to six feet high. During storms they can reach a much greater height. Slightly more protected sandy beaches occur at the mouth of Grays Harbor and Willapa bay, along the north side of the Olympic Peninsula, and on the exposed sides of San Juan and Whidbey islands.

Protected sand beaches are scattered throughout the inland waters of the state. They occur mainly at lower tidal levels and near the mouths of bays or rivers. Waves on protected beaches are usually less than one foot in height. Specific examples are found on the west side of Hood Canal, north of Port Gamble, along King County shorelines, and at Birch Bay in Whatcom County. It is common to find some gravel mixed in with the sand at the upper tidal levels.

Sand beaches, especially those in exposed situations, provide a highly stressful environment for several reasons. Sand particles are not capable of holding much moisture when the tide is out and therefore give organisms little protection against drying out. Sand particles also hold little detritus (decomposing organic matter) and few nutrients. Sand is subject to movement by wind, waves, and currents which provide a

Sand Beach Habitat

1) sandpiper
2) sand fleas
3) sea star
4) clam
5) starry flounder
6) heart cockle
7) shrimp
8) sand dollar
9) red rock crab
10) burrowing sea cucumber
11) sand sole
12) juvenile salmon

constantly shifting substrate. In addition, the beaches are usually accreting (building up) or eroding; therefore plants and animals are constantly being buried or uncovered.

As a result, the diversity and abundance of plants and animals on sand beaches is low compared to other beach types. Vegetation is restricted to benthic (bottom dwelling) or interstitial (occuring in small protected places) microscopic plants, as there is no stable substrate to which larger plants can attach.

The splash zone of exposed sand beaches usually consists of extensive sand dunes. (See Sand Dunes). Along less exposed beaches, the splash zone typically forms a narrow strip of beach grassland and often driftwood accumulates. The invertebrate most conspicuous in the high intertidal zone of exposed sand beaches is the beach hopper, or sand flea.

The mid-intertidal zones of exposed beaches are marked by the presence of filter feeding razor clams. Razor clams are one of the most abundant organisms on exposed beaches and have a great impact on the total number of animals present. Biomass of exposed sand beaches is very low when razor clams are absent. In the past few years the razor clam population has dropped significantly and razor clam digging is currently limited.

Also found along the mid-intertidal zone are worms, including the blood worm and the ribbon worm, which often reach up to 30 cm in length. Of these only blood worms inhabit beaches that are exposed to the full force of the Pacific Ocean. Ribbon worms and lugworms, however, occur in the mid-intertidal zones of protected beaches. The lugworm forms characteristic fecal mounds on the sediment surface which signal its presence. Burrowing species such as the lugworm are responsible for turning over marine sediments much as earthworms do in a garden. Small crustaceans also live here and are important food sources for bottom feeding fishes. The only large clam of these beaches is the white sand clam.

Two species related to sea urchins and starfish are common at times on protected sand beaches. Sand dollars, which feed on detritus, can become so dense that they form a strip along the shore which is visible in aerial photographs. Sand dollars also occur on exposed beaches in shallow subtidal areas. Much less conspicuous when present is the burrowing sea cucumber. Despite its name, it does not actually burrow beneath the sand, but lies partially buried on top. Both these species are widespread in Hood Canal and Puget Sound.

Virtually all of the species present in the mid-intertidal zone are present to a greater degree in the lower-intertidal zone, where several additional species also occur. The beautiful purple olive snail is restricted to the lower zones of exposed beaches on the outer coast. Protected beaches are dominated by a variety of worms and small crustaceans. During high tide the Dungeness crab, red rock crab, and a variety of shrimp invade the intertidal area to forage for food.

The number of fish species associated with sand beaches is relatively low compared to other beach types. Demersal fish (bottom dwellers) may be abundant at either exposed or protected areas and are often regarded as characteristic inhabitants of sandy nearshore waters. Representative species include English sole, sand sole, rock sole, and the Pacific staghorn sculpin.

The most characteristic wildlife inhabiting sandy beaches are sandpipers and the most typical species is the sanderling. Several species of shorebirds will often rest together in mixed flocks on the upper beach, sand spits, or in beach grass.

Sand beaches are extremely popular for recreation. They are ideal areas for walking, fishing, swimming, and sunbathing. Where eelgrass is present, Dungeness and red rock crabs can be found at low tide. There is also the chance to observe seals, sea lions, sea otters, and grey whales off the sand beaches of the outer coast.

The major impact by visitors on sand beaches comes from the collection of organisms such as sand dollars and moon snails as souvenirs. In addition, use of motor vehicles or just walking can damage organisms living at or near the sediment surface. Both these activities can kill resident species and reduce the food available for the birds and fish who feed here. As with all natural environments, care should be taken by the visitor not to harm the very things they enjoy.

ACKNOWLEDGEMENTS

The authors of this guide wish to acknowledge the contributions of many people who without their help and inspiration completion of this project would not have been possible.

The guide was originally conceived by John Black, a Seattle Attorney and Environmental Activist, after seeing a similar publication by the State of California. Glen Crandal of the State Department of Ecology became involved when funding from the National Oceanic and Atmospheric Administration's Coastal Zone Program was obtained. Initial work on the guide was accomplished by Dr. Gil Peterson, Professor of Geography at Western Washington University, and his students. Although much of this original work has been superceded the inspiration for the guide and the early drafts proved valuable in the preparation of this edition.

The initial source of data for the sites in the guide was an inventory of recreation sites maintained by the State of Washington Interagency Committee for Outdoor Recreation. This information was expanded and refined with data obtained from other sources and ultimately by field visits of virtually every site. Invaluable assistance was obtained from the many site owners, to numerous to list here, who reviewed site data and text to insure accuracy. A final and very helpful editing of the text was done by Lane Morgan.

The photographs were taken by the authors except where noted. Line drawings and graphics were done by Melly Reuling except for those completed by Deanna Hofmann. Computer data input was done by Linda Berlin who was temporarily hired for this seemingly never ending task.

We wish to credit our supervisors at the Department of Ecology. They exhibited great patience, under standing and confidence in us as this project grew in magnitude from its early conception to the several year undertaking it became.

ABOUT THE AUTHORS

JAMES W. SCOTT was the Public Shore Guide project leader. His chief contribution was organizing and preparing the site data for the guide. While envious associates sweltered in the stuffy office over more mundane state business, Jim conducted site investigations, took photographs and drove seemingly endless dead end roads looking for elusive public access sites. Before heading the public access program, he worked a number of years in various outdoor recreation jobs for the state. Jim graduated from Oregon State University and the University of Washington where he studied forest management and outdoor recreation.

MELLY A. REULING was hired as the chief research writer for this guide but in the course of its production found herself enjoying extensive field research and spending long evenings taking photos. She also did most of the artwork as well as the graphics design and layout for the guide. Melly is a recent graduate of the Evergreen State College where she studied environmental science and writing. When not working on the guide, Melly occupied much of her free time rock and mountain climbing and sea kayaking.

DON BALES was hired temporarily to draw the many maps in the guide, a job he performed with admirable skill and dedication. Don has a degree in geography/cartography from the University of Washington. His primary recreation is camping and observing nature.

SELECTED REFERENCES

Angell, Tony, *Marine Birds and Mammals of Puget Sound*, University of Washington Press, Seattle, Wa., 1982.

Carefoot, Thomas, *Pacific Seashores: A Guide to Intertidal Ecology*, University of Washington Press, Seattle, Wa., 1977.

Carson, Rachel L., *The Edge of the Sea*, Houghton Mifflin, Boston, Ma., 1955.

Downing, John, *The Coast of Puget Sound: Its Processes and Development*, University of Washington Press, Seattle, Wa., 1983.

An Ecological Characterization of the Pacific Northwest Coastal Region, 5 Vol., U.S. department of the Interior, Fish and Wildlife Service, Portland, Or., 1980.

The Ecology of Pacific Northwest Coastal Sand Dunes: A Community Profile, U.S. Department of the Interior, Fish and Wildlife Service, Portland, Or., 1984.

The Ecology of Tidal Marshes of the Pacific Northwest Coast: A Community Profile, U.S. Department of the Interior, Fish and Wildlife Service, Portland, Or., 1983.

Eltringham, S.K., *Life in Mud and Sand*, Crane, Russak, New York, 1971.

Estuaries: A Resource Worth Saving, Washington Department of Game, Olympia, Wa., 1972.

Hardy, A.C., *The Open Sea: Its Natural History*, Houghton Mifflin, Boston, Ma., 1965.

Hewlett, Stefani, *Sea Life of the Pacific Northwest*, Mcgraw-Hill Ryerson, New York, 1976.

Island County Shoreline Access Study, Island County Planning Department, Coupeville, Washington, 1977.

Johansen, D.O., and Gates, C.M., *Empire of the Columbia*, Harper and Row, New York, 1967.

Kozloff, Eugene N., *Seashore Life of Puget Sound, the Strait of Georgia, and the San Juan Archipelago*, University of Washington Press, Seattle, Wa., 1973.

Kozloff, Eugene N., *Seashore Life of the Northern Pacific Coast: An Illustrated Guide to Northern California, Oregon, Washington, and British Columbia*, University of Washington Press, Seattle, Wa., 1983.

Land Cover/Land Use Narratives, Washington State Department of Ecology, Olympia, Wa., 1980.

Marine Shoreline Study of Public Access and Recreation Sites in Whatcom County, Whatcom County Department of Parks and Recreation, Bellingham, Wa., 1976.

Nishitani, Louisa and Kenneth K. Chew, *Gathering Safe Shellfish in Washington: Avoiding Paralytic Shellfish Poisoning*, University of Washington Press, Seattle, Wa., 1982.

Pelz, Ruth, *The Washington Story: A History of Our State*, Seattle Public Schools, Seattle, Wa., 1977.

Perkins, Eric John, *The Biology of Estuaries and Coastal Waters*, Academic Press, London; New York, 1974.

Ricketts, Edward F., Jack Calvin and Joel W. Hedgpeth, *Between Pacific Tides*, 4th ed., Stanford University Press, Stanford, Ca., 1973.

Ruotsala, Andrew A., *Beach Processes in the Pacific Northwest*, Andrew A. Ruotsala, Seattle, 1979.

Skagit County Shoreline Access Study, Skagit County Planning Department, Mount Vernon, Wa., 1978.

Smith, Lynwood S., *Living Shores of the Pacific Northwest*, Pacific Search Press, Seattle, Wa., 1976.

Snohomish County Comprehensive Park and Recreation Plan, Snohomish County Parks and Recreation, Everett, Wa., 1984.

Speidel, Bill, *The Wet Side of the Mountains: Prowling Western Washington*, Nettle Creek Publishing Company, Seattle, Wa., 1974.

Strickland, Richard M., *The Fertile Fjord: Plankton in Puget Sound*, University of Washington Press, Seattle, Wa., 1983.

Teal, John and Mildred Teal, *Life and Death of the Salt Marsh*, Little, Brown, Boston, 1969.

Washington State Coastal Zone Management Program, Washington State Department of Ecology, Olympia, Wa., 1976.

Younge, Charles Maurice, *The Seashore*, Revised ed. Collins, New York, 1966.

Your Public Beaches, a series of five guides to Puget Sound public tidelands, Washington Department of Natural Resources, Olympia, Wa., 1975-1978.

INDEX

10th Street Beach Access, 319
1st Ave. South Bridge Boat Ramp, 169
20th Place S.W. Street End Access, 173
36th N.W. Street End Boat Launch, 213

Aberdeen, 290
Access Sign, 9
Acknowledgements, 338
Admiralty Bay, Beach 124, 133
Admiralty Bay, Beach 124A, 133
Agate Bay, Beach 420, 257
Agate Bay, Beach 421, 257
Agate Beach County Park, 107
Alaska Square, 167
Alek Bay, Beach 308, 109
Alki Beach Park, 165
Alki Point Light Station, 165
Allyn Park, 234
Allyn Port and Dock, 237
Altoona, 323
American Camp, 97
American Legion Park, 191
Anacortes, 45
Anacortes Ferry Terminal, 55
Anderson Island, Beach 8, 221
Andover Place, 169
Anna Smith Park, 191
Annapolis Public Access Area, 201
Anthony's Home Port, Public Access, 161
Arcadia Launching Ramp, 245
Armitage Island, Beach 290A, 111
Arness County Park, 187
Asian expeditions, 16

Bachmann Park, 193
Ballard Elks Public Access, 161
Barnes Island, Beach 229, 85
Bayview Boat Launch, 47
Bayview Market Public Access, 229
Bayview State Park, 47
Beach 1 (Olympic National Park), 275
Beach 2 (Olympic National Park), 275

Beach 3 (Olympic National Park), 275
Beach 4 (Olympic National Park), 275
Beach 6 (Olympic National Park), 275
Beach Haven, Beach 238, 77
Beach Hiking, 272
Beaches, 326
Beard's Hollow, 321
Belfair State Park, 237
Bellingham, 31
Berg Drive N.W. Boat Launch, 215
Big Ditch Access, 149
Birch Bay County Park, 33
Birch Bay State Park, 33
Birch Point, Beach 372, 33
Blaine Harbor and Boat Launch, 33
Blake Island State Park, 197
Blakely Island, Beach 292, 111
Blakely Island, Beach 292A, 111
Blind Bay, Beach 260D, 117
Blind Island State Park, 117
Bolstead Beach Access, 319
Bolton Peninsula, Beach 56, 285
Bonge Ave. Beach Access, 295
Boston Harbor Boat Ramp, 231
Boulevard Park, 39
Bowerman Basin, 290, 299, 302
Bracketts Landing Beach, 155
Breazeale-Padilla Bay Interpretive Center, 48
Bremerton, 183
Bremerton Ferry Terminal, 201
British Empire, 16
Broad Spit, 285
Broken Point, Beach 260A, 119
Browns Point Lighthouse Park, 207
Bruceport County Park, 313
Bruceport Historical Marker, 313
Bumstead Spit South, Beach 223A, 37
Bumstead Spit, Beach 223, 37
Burfoot County Park, 231
Burrows Island State Park, 55
Burton Acres County Park, 181
Bush Pioneer County Park, 313

Bush Point, Beach 101, 135
Butter clams, 247
Bywater Bay State Park, 285

Cactus Islands, Beach 353A, 89
Cactus Islands, Beach 353B, 89
Camano Island, 122
Camano Island State Park, 141
Camping, 22
Canoe Island, Beach 296A, 117
Cap Sante Boat Haven, 57
Cap Sante Park, 57
Cape George, Beach 407, 283
Cape St. Mary, Beach 311, 109
Capitol, State, 227
Captain Robert Gray, 16
Captain George Vancouver, 16
Carkeek Park, 161
Carrs Landing Public Access, 209
Carter Point, 41
Case Shoal, Beach 59A, 285
Castle Island State Park, 109
Cathlamet, 323
Catlina Shores Marine Park, 127
Cattle Point Lighthouse Recreation Site, 97
Cavalero Beach County Park, 141
Center Island Recreation Site, 115
Center Island, Beach 324A, 115
Central Floats, 39
Chance A La Mer Beach Access, 297
Chehalis River, 290
Chetzemoka Park, 281
Chinook County Park, 321
Chuckanut Bay Park, 39
City Beach Park, 127
City Park, 211
City Pier, 269
City Waterway Dock, 207
Clallam Bay State Park, 259
Clallam County, 250
Clam Digging, 246
Clark Island State Park, 85

Cline Spit County Park, 265
Clinton Ferry Terminal, 139
Clinton Recreational Pier, 139
Clyde V. Davidson Fishing Pier, 221
Coast Guard Museum, 167
Coastal Indians, 14, 15
Cockle, 248
Coleman Dock, 167
Columbia River forts, 145
Columbia White-tailed Deer Nat'l Wildlife Ref., 323, 325
Commencement Park, 209
Commodore Park, 163
Cone Island State Park, 51
Conservation, 12, 18
Coon Island, Beach 245A, 121
Copalis Beach Access, 293
Cottonwood Beach County Park, 33
County Line Park, 325
Coupeville Wharf, 129
Crabbing, 198
Cranberry Beach Access, 315
Crane Island, Beach 250A, 121
Crane Island, Beach 250B, 121
Crystal Springs Public Fishing Pier, 195
Cutts Island State Park, 213
Cypress Head Recreation Site, 53
Cypress Head, Beach 211, 53
Cypress Island, Beach 209, 53
Cypress Island, Beach 210, 53

Darlington Beach/Tidelands, 153
Dash Point Park, 207
Dash Point State Park, 173
Dave Mackie Memorial County Park, 139
Davis Slough, 149
Davis Slough Access Area, 149
Decatur Beach, Beach 324, 113
Decatur Head/White Cliff, Beach 323, 113
Decatur Island, Beach 319A, 115
Deception Pass State Park, 65, 123, 125
Deep Creek, 259
Deer Harbor, Beach 240B, 81
Deer Point, Beach 277, 69
Department of Ecology, 1, 9, 11, 48
Department of Game, 25
Department of Natural Resources, 21

Department of Natural Resources Beaches, 21
Department of Social and Health Services, 42
Des Moines Fishing Pier, 171
Des Moines Marina, 171
Devils Head, Beach 13, 225
Devils Slide, Beach 220A, 41
Dewatto Bay, Beach 44A, 241
Diamond Point, Beach 265, 77
Diamond Point, Beach 410, 261
Dikes, 62
Dinoflagellates, 42
Discovery Park, 163
Dockton County Park, 181
Doe Bay, Beach 281A, 73
Doe Island State Park, 73
Don Armeni Park, 165
Dosewallips State Park, 287
Double Bluff East Beach, 137
Double Island, Beach 251, 81
Double Island, Beach 251A, 81
Draining, 62
Dredging, 62
Dry Creek, Beach 414, 269
Dugualla Bay County Park, 125
Dugualla Bay, Beach 142, 127
Dugualla Bay, Beach 144, 127
Dumas Bay Park Wildlife Sanctuary, 173
Dune Plant Habitat, 307
Dunes, 306
Dungeness Boat Launch, 265
Dungeness crab, 198
Dungeness National Wildlife Refuge, 265, 266
Dungeness Recreation Area, 265, 266
Dungeness Spit, 250, 266
Duwamish Head, 165
Duwamish Waterway Park, 169

Eagle Cliff, Beach 286, 51
Eagle Cove County Park, 97
Eagle Harbor Park, 195
Eagle Harbor, Beach 212A, 51
Eagle Island State Park, 221
East Beach County Park, 279
East Boat Haven Boat Launch, 269
East Sound, Beach 266, 77
East Sound, Beach 267, 77
East Sound, Beach 270, 77

East Sound, Beach 274, 75
East Sound, Beach 275, 75
East Vashon County Park, 175
Ebey's Landing State Park, 131
Ecology, Department of, 1, 9, 11, 48
Ed Munro/Seahurst County Park, 171
Ediz Hook Boat Launch, 269
Edmonds Marina Beach, 155
Eglon Boat Launch, 185
Elephant seal, 34
Elliott Bay Park, 167
Elochoman Slough Marina, 325
Emma Schmitz Memorial, 169
End of 146th Ave. S.W., 179
End of 182nd Ave., KPN, 225
End of 37th Street N.W. Boat Launch, 215
End of 9th Ave., Fox Island, 215
End of Admiralty Ave., 135
End of Cultus Bay Road, 139
End of Kamas Drive, Fox Island, 213
End of Main Street, 135
End of Olman Road, KPN, 223
End of Point Fosdick Road, 215
End of Soundview Drive N.W., 149
English Camp Historical Park, 95
Estuaries, 60
Estuarine habitats, 60
Estuarine Plant Habitat, 63
Everett Jetty State Park, 151
Evergreen Park, 201
Ewing Island, Beach 367A, 83

Fay Bainbridge State Park, 189
Ferries, 176
Ferry Routes, 177
Fire Station No. 5, 209
Fireman's Park, 207
First Street Dock, 201
Fish Point Park, 37
Fisheries, 7
Fisherman Bay, Beach 299, 101
Fishermans Terminal, 163
Fishery Point, Beach 363, 87
Fishing Bay, Beach 270A, 77
Flapjack Cove Tidelands, Beach 54, 287
Flat Point, Beach 295, 101
Flintstone Park, 127

Floating Docks, 156
Floats, 156
Flower Isle, Beach 266B, 77
Forest Park, 151
Fort Canby, 145, 305
Fort Canby State Park, 321
Fort Casey, 143
Fort Casey State Park, 133
Fort Columbia, 145, 305
Fort Columbia Historical State Park, 321
Fort Ebey, 143
Fort Ebey State Park, 131
Fort Flagler, 142, 270
Fort Flagler State Park, 281
Fort Hayden, 143
Fort Ward, 143
Fort Ward State Park, 195
Fort Worden, 142, 270
Fort Worden State Park, 283
Forts, 142
Foulweather Bluff, Beach 64, 185
Franciscan Dock, 117
Freeland County Park, 137
Freeman Island State Park, 79
Freshwater Bay Boat Launch, 269
Freshwater Bay, Beach 416, 269
Freshwater Bay, Beach 417, 269
Frost Island, Beach 318, 99
Frye Cove County Park, 229
Fudge Point, Beach 24, 243
Fur trade, 16

Game, Department of, 25
Gardiner Public Boat Launch, 283
Geoducks, 249
Gibson Spit, Beach 411, 261
Glacial deposits, 14
Glacial Ice Age, 14
Glendale Road End, 139
Glendale, Beach 100, 139
Glendale, Beach 99, 139
Golden Gardens Park, 161
Gonyaulax catenella, 42
Grapeview Boat Ramp, 237
Gravel beach, 329
Gravel beach habitat, 330
Grayland Beach Access, 295

Grayland Beach State Park, 311
Grays Harbor, 62, 290, 291
Grays Harbor County, 290
Grays Harbor Estuary, 302
Griffin Bay Recreation Site, 97
Griffin Bay, Beach 326, 97
Griffith Priday State Park, 294
Guemes Island, Beach 199C, 55

H.J. Carroll State Park, 287
Habitat Management Areas, 25
Hadlock Boat Launch, 279
Hall Road Street End Boat Launch, 223
Hamilton Park, 209
Hamilton Viewpoint Park, 165
Hanikin Point, Beach 264, 117
Hansville, Beach 69, 185
Harbor Island Marina, 161
Harbor seal, 34
Harbor Vista Park, 165
Harborview Park, 153
Hardshell clams, 247
Harney Channel, Beach 262, 77
Harper County Park, 197
Harper Public Fishing Pier, 197
Harstene Bridge Boat Ramp, 243
Harstene Island, Beach 33, 243
Harvey Rendsland State Park, 239
Hastie Lake Road Boat Launch, 131
Henry Island, Beach 339A, 95
Hicks County Park, 285
Highway 112 West of Sekiu River, 253
Hiram M. Chittenden Locks, 163
History, 14
History of Public Beaches, 20
Hoh River, 276
Hoko River, Beach 428, 253
Home Boat Launch, 223
Hood Canal, 234
Hood Canal Recreation Park, 239
Hood Canal Salmon Hatchery, 239
Hood Canal, Beach 46, 241
Hood Canal, Beach 47, 241
Hood Canal, Beach 48, 241
Hoodsport, 235
Hoodsport, Beach 43, 239
Hope Island State Park, 65

Horse clams, 247
Howarth Park, 151
Hudson's Bay Company, 17
Hunter Bay Dock County Park, 103
Hunter Bay, Beach 313, 103
Hunter Bay, Beach 313A, 103
Hunter Bay, Beach 314, 103
Hurricane Ridge, 250

Iceberg Island State Park, 107
Illahee Pier, 193
Ilwaco Marina, 321
Indian Cove, Beach 296, 117
Indian Trail, 171
Indianola Dock, 187
Intertidal zonation, 326
Island County, 122

J.B. Pope Marine Park, 281
Jackson Cove, Beach 55, 287
James Island State Park, 113
Jarrell Cove State Park, 243
Jarrell Cove, Beach 34, 243
Jefferson County, 270
Jensen Access, 59
Jerisich Park, 211
John Wayne Marina, 261
John's River Habitat Management Area, 290, 299
Johns Island, 89
Johns Island, Beach 356, 89
Johns Point, Beach 307, 107
Jones Island State Park, 121
Jorsted Creek Beach, 241
Joseph Whidbey State Park, 127

Kalaloch Campground, 275
Kayak Point County Park, 149
Kellett Bluff, Beach 341, 94
Key to site information, 26
Keyport County Park, 189
Keystone Beach Tidelands, 133
Keystone Ferry Terminal, 133
Killer whales, 104
King County, 159
Kingston Ferry Terminal, 187
Kingston Marina, 187
Kinney Point, Beach 404A, 279

Kitsap County, 182
Kitsap Memorial State Park, 187
Klipsan Beach Access, 315
Kopachuck State Park, 213

L.B. Good Memorial Park, 279
La Push Beach #1, 255
La Push Beach #2, 255
La Push Beach #3, 275
La Push Marina, 255
LaConner, 45
LaConner Marina, 59
LaConner Waterfront, 59
Lake Ozette Ranger Station, 273
Larrabee State Park, 41, 47
Leadbetter Point, 305, 316
Leadbetter Point State Park, 315
Leadbetter Point Unit, Willapa Nat'l Wildlife Ref., 313
League Island Access, 149
Legislature, 10
Lents Lane, 193
Lewis and Clark Campsite State Park, 321
Lewis Unit, Willapa National Wildlife Refuge, 319
Libbey Beach County Park, 131
Liberty Bay Park, 191
Lighthouse Marine County Park, 33
Lilliwaup Public Beach, 241
Lilliwaup Tidelands State Park, 241
Lime Kiln Point County Park, 97
Lincoln Park, 169
Lions Community Playfield, 193
Little Patos Island, Beach 366A, 83
Little Sucia Island, Beach 367D, 83
Littleneck clams, 247
Long Beach Peninsula, 305, 306
Long Island, 304, 316
Long Island Unit, Willapa Nat'l Wildlife Refuge, 315
Long Point Beach, 129
Longbranch Boat Launch, 225
Longbranch Dock, 225
Loomis Lake State Park, 315
Lopez Ferry Terminal, 99
Lopez Island, Beach 305, 107
Lopez Pass, Beach 312A, 103
Lopez Sound, Beach 315, 103
Lopez Sound, Beach 317, 99
Lover's Cove, Beach 239, 79
Lowman Beach Park, 169
Lummi Island Recreation Site, 41
Lummi Island, Beach 220, 41
Lummi Island, Beach 223B, 37
Lummi Island, Beach 224, 37
Lummi Peninsula, 31

Mackaye Harbor, Beach 306, 107
Magnolia Park, 163
Manchester Boat Launch, 197
Manchester State Park, 195
Manila clams, 247
Maple Grove Boat Launch, 141
Maple Hollow Recreation Site, 223
March Point Recreational Beach, 57
March Point Tidelands, 57
Marine facilities, 23
Marine Mammal Protection Act, 13, 34
Marine mammals, 7, 104
Marine Park, 209
Marine Park Boat Launch, 151
Mariner's Cove Boat Launch, 127
Mason County, 234
Matia Island State Park, 85
Mats Mats Bay Boat Launch, 285
Maury Island, Beach 83, 181
McArdle Bay, Beach 309, 109
McConnell Island, Beach 245, 121
McCracken Point, Beach 340, 95
McMicken Island State Park, 243
Me-Kwa Mooks Park, 169
Meadowdale County Park, 155
Memorial Park, 207
Midway Beach Access, 311
Migratory Birds Treaty Act, 300
Milltown Access, 59
Mini Harbor Park, 127
Mission Beach Park, 149
Mitchell Bay Islet, 95
Moclips Beach Access, 293
Moclips Cetological Society, 104
Monroe's Landing County Park, 129
Monument County Park, 33
Mora Ranger Station, 273
Moran State Park, 73

Moran's Beach County Park, 125
Mosquito Pass, Beach 344, 95
Mountain View Road End, 135
Mud Bay, Beach P1, 107
Mud Bay, Beach P2, 107
Mud beach, 332
Mud beach habitat, 333
Mukilteo Fishing Pier, 153
Mukilteo State Park, 153
Mussels, 248
Mutiny Bay Boat Launch, 137
Myrtle Edwards Park, 167
Mystery Bay State Park, 279

N.A.D. Marine Park, 193
N.E. Vashon County Park, 175
Nahcotta Small Boat Basin, 315
Nakeeta Beach/Tidelands, 153
National Estuarine Sanctuary, 48
National Park Service, 272
National Wildlife Refuges, 24, 233
Native Americans, 14
Natural Resources, Department of, 21
Neah Bay, 250
Neah Bay Picnic Area, 253
Neap tide, 218
Near The Herron Ferry Terminal, 225
Neck Point, Beach 259A, 119
Nisqually Habitat Management Area, 231
Nisqually National Wildlife Refuge, 231, 233
NOAA weather advisory, 272
Normandy Beach Park, 171
North Beach County Park, 283
North Beach Road End, 79
North End Of Driftwood Way, 133
North Finger Island, Beach 367B, 83
North Fork Access, 59
North Jetty Beach Access, 299
North Marine View Park, 151
North River Public Fishing Site, 311
Northcove Beach Access, 311
Northeast Stuart Island, Beach 356A, 91
Northwest Coast Indians, 14, 15
Northwest Island Marine Park, 65
Northwest McConnell Rock State Park, 121

Oak Bay County Park, 279

Oak Bay Sand Spit, 279
Oak Island, Beach 257A, 81
Obstruction Pass Boat Launch, 69
Obstruction Pass County Park, 69
Obstruction Pass Recreation Site, 69
Obstruction Pass, Beach 276, 69
Ocean City Beach Access, 293
Ocean City State Park, 297
Ocean Lake Way Beach Access, 297
Ocean Park Beach Access, 315
Ocean Shores Interpretive Center, 299
Odlin County Park, 99
Old Fort Townsend State Park, 281
Old Hat Slough Bridge, 149
Old Man House State Park, 189
Old Town Dock, 209
Olga County Park, 75
Olga Marine State Park, 75
Ollala Boat Launch, 197
Olympia, 227
Olympia Isle Marina, 229
Olympic Beach Park, 155
Olympic National Forest, 235, 276
Olympic National Park, 250, 270, 276
Olympic Peninsula, 6
Orcas, 104
Orcas Island Ferry Terminal Picnic Area, 81
Orcas Island, Beach 279, 69
Orcas Island, Beach 282, 73
Orcas Island, Beach 283, 73
Oregon Country, 17
Oyhut Beach Access, 297
Oyhut Habitat Management Area, 299
Oysters, 248
Oysterville Beach Access, 315
Ozette Beach Access, 255

Pacific Beach State Park, 293
Pacific Coast, 6, 250, 270
Pacific County, 304, 316
Pacific Flyway, 7
Pacific Pines State Park, 315
Pacific Way Beach Access, 297
Padilla Bay, 45, 48
Padilla Bay National Estuarine Sanctuary, 47
Paralytic shellfish poisoning, 42, 246
Park Ave. Street End, 153

Patos Island State Park, 83
Pear Point, Beach 332, 93
Pelican Beach Recreation Site, 51
Penn Cove on Deception Pass, 129
Penn Cove Tidelands, 129
Penrose Point State Park, 223
Percival Landing, 229
Percival Landing North, 229
Phil Simon Memorial Park, 137
Picnic Point County Park, 155
Pier 48 Viewpoint, 167
Pierce County, 205
Pilings, 156
Pillar Point County Park, 259
Pillar Point, Beach 424, 259
Pillar Point, Beach 425, 259
Pioneer Orchard Park, 221
Pioneer Park, 59
Pit Passage, Beach 6, 225
Pleasant Harbor State Park, 283
Point Colville, 109
Point Defiance City Park, 205
Point Defiance Park, 207
Point Defiance Zoo, 205
Point Doughty Recreation Site, 79
Point Doughty, Beach 236, 79
Point Evans, Beach 36, 211
Point Fosdick, Beach 1, 215
Point Fosdick, Beach 1A, 215
Point Hammond, Beach 362, 87
Point Lawrence Recreation Site, 73
Point Lawrence, Beach 231, 73
Point No Point County Park, 185
Point No Point, Beach 68, 185
Point Partridge Recreation Site, 131
Point Robinson County Park, 181
Point Thompson, Beach 234, 75
Point Whitney Tidelands, 289
Port Angeles, 250
Port Gamble, 183
Port of Allyn Public Boat Ramp, 237
Port of Brownsville Marine Park and Marina, 193
Port of Friday Harbor, 93
Port Orchard Boat Launch, 201
Port Orchard Marina, 201
Port Townsend, 270
Port Townsend Boat Basin, 281

Port Townsend North Pier, 281
Port Williams Boat Launch, 261
Portage Island, 31
Posey Island State Park, 95
Potlatch State Park, 239
Poulsbo, 183
Poulsbo Boat Launch and Marina, 191
Poverty Bay County Park, 173
Precipitation, 6
President's Channel, Beach 240, 79
Priest Point Park, 229
Protection Island, 270
PSP, 42, 246
Public access, 2
Public Access Logo, 9
Public beaches, 20
Public Fishing and Boat Launch, 313
Public Viewpoint, 155
Puget Sound, 6, 17
Puget Sound forts, 142

Quilcene Boat Haven, 285
Quinalt River, 276

Raccoon Point, Beach 233, 75
Rain Forests, 276
Ram Island, Beach 312B, 103
Randall Drive Boat Launch, 211
Razor clams, 249
Reads Bay, Beach 319, 115
Reads Bay, Beach 325, 113
Red rock crab, 198
Red Tide, 42, 246
Red tide hotline, 42
Redondo County Park, 173
Refuges, 24
Resource management, 18
Restoration Point, 195
Rialto Beach, 255
Richmond Beach County Park, 161
Riekkola Unit, Willapa Nat'l Wildlife Refuge, 319
Right Smart Cove State Park, 287
Road End of Spur Road Off Borgman Road, 127
Roadside Rest Area, 321
Roadside Viewpoint 'Big Wheel', 245
Robert F. Kennedy Recreation Site, 225
Rock beach, 327

Rock beach habitat, 328
Rock Point, Beach 303, 101
Rocky Bay/Limestone Point, Beach 336, 93
Roosevelt Beach Access, 293
Rosario, Beach 272, 75
Ross Point Tidelands, 197
Ruby Beach, 275
Rueben Tarte County Park, 93
Russian expedition, 16

Saddlebag Island State Park, 47
Salisbury Point County Park, 185
Salmon Road End, 135
Salt Creek Recreation Area, 257
Salters Point Beach, 221
Saltwater State Park, 173
Samish Island Recreation Site, 47
San Juan Channel, Beach 298, 101
San Juan Channel, Beach 334, 93
San Juan County, 66
San Juan County Park, 95
San Juan Island, Beach 330, 93
San Juan National Wildlife Refuge, 66, 67, 69, 70
San Juans, 66
Sand beach, 334
Sand beach habitat, 335
Sand dunes, 306
Sandy Point, Beach 364, 87
Saratoga Pass Tidelands, 133
Satellite Island, Beach 358, 91
Scenic Beach State Park, 203
Seabirds, 70
Seal Rock Campground, 287
Seals, 34
Seashore Conservation Act, 22
Seattle, 159
Seattle Aquarium, 159
Seaview Ocean Beach Access, 319
Seawall Park, 137
Sekiu Point, Beach 427, 253
Sekiu Public Area, 253
Sekiu River Access, 253
Sekiu River, Beach 429A, 253
Semiahmoo County Park, 33
Sequim Bay Boat Launch, 261
Sequim Bay State Park, 261
Shark Reef Recreation Site, 101

Shark Reef, Beach 304, 101
Sharpe County Park, 65
Shaw County Park, 117
Shaw Island, Beach 258, 119
Shaw Island, Beach 260, 119
Shaw Island, Beach 260C, 117
Sheep Island, Beach 255A, 81
Shellfish, 7, 247
Shelton, 234
Shelton Boat Ramp, 245
Shenanigan's Public Access, 209
Shi Shi Beach, 255
Shilshole Bay Marina, 161
Shine Tidelands, 285
Shipwreck Point, Beach 429, 253
Shorebirds, 300
Shorecrest County Park, 245
Shorelands Program, 9
Shoreline management, 10
Shoreline Management Act, 1, 10
Silverdale County Park, 191
Sinclair Island Dock, 51
Sinclair Island Light, Beach 213A, 51
Sinclair Island, Beach 213, 51
Skagit County, 45
Skagit Habitat Management Area, 45, 59
Skagit Island State Park, 65
Skagit River estuary, 147
Skomokawa, 323
Skomokawa Vista Park, 325
Skull Island State Park, 81
Slip Point, Beach 426, 259
Smith Cove Park, 165
Smugglers Cove North, Beach 221A, 41
Smugglers Cove Point, Beach 221, 41
Snohomish County, 147
Snohomish River Public Boat Launch, 151
Snow Creek Boat Launch, 253
Softshell clams, 249
South Beach Camp Area, 275
South East Vashon Island, Beach 79, 179
South Finger Island, Beach 367C, 83
South Indian Island County Park, 279
South Marine View Park, 151
South Shore Drive Road End, 55
South Side Boat Ramp, 39
South Terminal Park, 39

South Whidbey State Park, 135
Southeast Stuart Island, Beach 356B, 91
Spencer Spit State Park, 99
Spieden Bluff, Beach 353, 89
Spieden Island, Beach 352, 89
Spieden Island, Beach 352A, 89
Spits, 262
Spring Beach County Park, 179
Spring Passage/North Pass, Beach 240A, 79
Spring Tide, 217
Squalicum Harbor, 39
Squamish Harbor, Beach 59, 285
Squaxin County Park, 245
Squaxin Island State Park, 245
State Capitol, 227
State Environmental Policy Act, 10
State Park rules, 22
State Parks, 22
State Shellfish Lab, 316
Steilacoom Boat Launch, 221
Strawberry (Loon) Island Recreation Site, 53
Strawberry Bay, Beach 287, 53
Street End Viewpoint, 211
Stretch Island, Beach 20, 237
Stretch Point State Park, 237
Striped Peak Recreation Area, 257
Striped Peak, Beach 419, 257
Stuart Island State Park, 91
Stuart Island, Beach 359, 91
Sucia Island State Park, 85
Sunnyside Beach Park, 221
Sunrise Beach Park, 211
Suquamish Center, 189
Suquamish Museum, 183
Suquamish Museum and Tribal Center, 189
Swinomish Channel Boat Launch, 65

Tabook Point, Beach 57, 289
Tacoma, 205
Taurus Street Beach Access, 297
Taylor Bay, Beach 16, 225
Teekalet Bluff, 185
Thatcher Pass, Beach 291, 111
Thatcher Pass/Fauntleroy Point, Beach 322, 113
Thompson Road Access Site, 65
Three Tree Point, 171
Three Tree Street Corner Access, 171

Thurston County, 227
Tidal Cycle, 219
Tidelands, 20
Tidepools, 31
Tides, 216
Titlow Park, 215
Toandos Peninsula, Beach 57B, 289
Toandos Tidelands State Park, 289
Tokeland Marina, 311
Tolmie State Park, 231
Towhead Island County Park, 213
Towhead Island, Beach 285, 51
Town Boat Launch, 129
Town Park, 129
Toxic shellfish, 43
Tracyton Boat Launch, 193
Tramp Harbor Fishing Pier, 179
Transportation, Department of, 176
Travis Spit, Beach 411A, 261
Triton Cove, Beach 50, 287
Trump Island, Beach 320, 115
Turn Island State Park, 93
Twanoh State Park, 237
Twin Harbors State Park, 295
Twin Rivers, Beach 422, 257
Twin Rivers, Beach 423, 259
Twin Rivers, Beach 423A, 257
Twin Rocks State Park, 77

U.S. Fish and Wildlife Service, 24

Union Public Launching Area, 239
University of Washington Marine Lab, 66
Unnamed Isle, Beach 266B, 77
Unnamed State Park, Beach 325A, 113
Upright Channel Recreation Site, 101
Upright Head, Beach 294, 99
Utsalady No.1 County Park, 141

Vashon Roadside Viewpoint, 179
Vaughn Bay Spit, Beach 18, 223
Vendovi Island, Beach 214, 47

Wahkiakum County, 323
Waldron Island, Beach 361, 87
Waldron Island, Beach 361A, 87
Walker County Park, 245
Washington Department of Ecology, 48
Washington Department of Game, 25
Washington Department of Transportation, 176
Washington forts, 142
Washington Park, 55
Washington State Ferries, 176
Washington State Parks and Recreation Commission, 22
Washington statehood, 18
Washington Street Boat Harbor, 167
Washington Territory, 17
Washington's Shoreline Management, 10
Wasp Passage, Beach 259, 119
Waterfront Park, 167

Wauna Boat Launch, 213
Wauna, Beach 35, 213
Wauna, Beach 35A, 213
Weather advisory, 272
West Boat Haven Boat Launch, 269
West Pass Bridge, 149
West Side Viewpoint, 197
West Vashon Island, Beach 77, 175
West Vashon Island, Beach 78, 175
Westhaven State Park, 295
Westport Light State Park, 295
Westport Marina, 299
Wetlands, 60
Whale Museum, 66, 104
Whatcom County, 31
Whidbey Island, 122
White Rock, 285
Willapa Bay, 60, 304
Willapa National Wildlife Refuge, 316
Willow Island, 111
Winslow Ferry Terminal, 195
Woodward Bay Road Bridge, 231
Wyckoff Shoal, Beach 39, 223
Wynn-Jones County Park, 195

Young County Park, 55